AF879859

Brenda van Rensburg

ON BOARD AI

The Executive's Guide to AI, Risk, Governance and Implementation.

© 2025 Terrene Global
All Rights Reserved

No part of this publication may be reproduced, stored in or introduced into a retrieval system, or transmitted, in any form, or by any means (electronic, mechanical, photocopying, recording, or otherwise), without the prior permission of the publisher or author.Requests for permission should be directed to tgpublisher@outlook.com.

The web addresses reference in this book were live and correct at the time of the book's publication, and are subject to change.

First Published in 2025 by
Terrene Gobal

Title: On Board AI : *The Executive's Guide to AI, Risk, Governance and Implementation.*

Author: Brenda van Rensburg
ISBN: 978-0-9923564-0-8

Cover Design: Gillian Taylor
Quotes: Brenda van Rensburg

First edition: 2025

ISBN: 978-0-9923564-0-8

10 9 8 7 6 5 4 3 2 1

Dedicated to my *loving husband* and
my two *amazing daughters* (Gillian & Cheri)

In *loving memory* of my late sister (Fiona) and
my late brother-in-law (Wikus)

Table of Contents

Table of Contents

“AI is like a
relationship.
It can begin
with promise
OR
end with regret.
The difference lies in
how it’s managed”

Introduction

The Journey of Implementing AI Responsibly

Artificial Intelligence (AI) is the new 'shiny toy' wrapped in FOMO bubble wrap. Most people want it, but very few know why. It has immense capability, but not without the risk. Handled well, AI can turn the ordinary into the extraordinary. Handled poorly, it can steer a business off course entirely. **The risks**, as I will discuss throughout this book, **are not only technical but strategic, ethical, and human**.

AI has moved from the margins of innovation to the center of decision-making. It now shapes how governments deliver services, businesses compete, and societies function. For directors and executives, this shift is no longer just a technical issue. It is a question of governance, risk, trust, and accountability. The choices made in boardrooms will determine whether AI becomes a force for progress or a source of systemic exposure.

This book was written to help boards and executives navigate that frontier. It brings together the language of technology with the discipline of governance, turning ideas into practical application. Questions executives and directors must ask:

- Who is accountable for the outcomes of AI?
- How do we ensure fairness, transparency, and human oversight?
- When does innovation cross into irresponsibility?

Artificial intelligence challenges the foundations of governance. Traditional frameworks were built for predictable risks such as financial, operational, or reputational concerns. AI brings probabilistic systems, new data dependencies, and opaque decision-making processes that change faster than policy can keep up. The result is a growing gap between what technology can do and how organizations control it. I have written this book to help close that gap, putting AI into structures that boards already know, such as strategy, governance, risk, and ethics.

My book is separated into four parts. The first is all about AI. I start by looking at what AI actually is and simplify it by exploring the whole and the parts. I further offer practical tools to navigate AI risks, governance and its implementation.

The second part begins with risk. It then moves to AI risk and introduces the Big Four, which sits at the center of securing AI. From there, I explain AI security risks and show how to reduce them, making the topic less confronting and more practical.

Part three is about governing AI. The discussion starts with responsibility, liability, and duty of care. I expand on different governance models grounding on elements of each that become features within AI

Governance. I move into ethics, discussing ESG principles. The core of this chapter addresses the responsible and ethical governance of AI.

Finally, the fourth part incorporates all the previous parts of the book to illustrate how to implement AI. It covers strategy, purpose, and scale. I close the chapter with two examples. One was successful and the other was not. Both these stories illustrate the when done correctly, AI can impact growth and culture.

The future of AI leadership will belong to those who combine technical understanding with clarity. Governance, risk, and implementation are not barriers to progress. They are the foundation to a future with AI. My goal is that this book gives boards and executives the tools to lead responsibly, the courage to challenge assumptions, and the vision to shape AI that reflects the best of human intention.

PART 1

Understanding AI

"When AI fails,
it is not because it
isn't human.
It fails because it
performs tasks that
are human-like
without
Comprehension."

CHAPTER 1

Artificial Intelligence (AI)

Many people treat AI as if it was one thing, assuming every system can do the same job. It is understandable to take this approach. After all, the term is everywhere. Vendors market products as "AI-powered." News headlines warn of AI gone wrong. Politicians debate regulation. Yet the word itself is often misunderstood.

AI is not human. It does not think or reason the way people do. But it can perform tasks with human-like intelligence, from recognizing patterns in medical images to generating fluent text. That distinction matters. If we confuse AI with human intelligence, we risk both overestimating and underestimating what it can do.

Artificial Intelligence, or AI, is often defined in textbooks as a branch of computer science focused on building machines that can perform tasks with "human-like intelligence."[1] That sounds straightforward, but very few people stop to ask what "human-like" really means. Is it the intelligence of a toddler? Is it tied to something measurable, like an IQ score? Or does it mean something entirely different?

The phrase "human-like" intelligence in AI can be misleading. It does not mean the system understands, or reasons, in the way we do. Instead, it means the system is able to perform tasks that, until recently, required some element of human judgment. These include tasks such as recognizing patterns in medical images, summarizing long legal contracts, or generating new text. This is why, when AI goes wrong, it makes headlines. The failure is not that it did not act "human," but that it was performing tasks that look human-like without any human comprehension.

Despite the "AI got it wrong" headlines, interest has only grown. In 2025, the word "AI" was typed into Google more than 2.2 million times.[2] Those searches didn't just come from coding engineers or students. They came from employees, regulators, executives, and consumers with interest to understand more about AI and how it could impact them. That search volume is a signal. AI has entered the public domain. It is no longer limited to researchers or big tech labs. It is a household term, and the expectations, together with concerns about what it can do, have found its way to the boardroom.

This chapter is about clarity. We will strip away the hype, separate AI from algorithms and automation, and explore what makes up an AI system. We will look at the difference between an AI model, the practices around it, and the full AI system that boards are responsible for.

A 'Bit' of History

So how did we get here? To understand what AI is today, it helps by understanding where it has come from. The story begins in the 1950s, when Alan Turing starts his paper 'Computing Machinery and Intelligence' with the line, "I propose to consider the question, Can machines think?"[3] The thought experiment that followed, known then as the Imitation Game and later as the Turing Test, became a landmark in the history of artificial intelligence.

Soon after, in 1956, the Dartmouth Conference[4] formally introduced the term artificial intelligence, marking the beginning of the field. In the 1960s, Joseph Weizenbaum created ELIZA[5], an early natural language program that mimicked a psychotherapist by reflecting statements back as questions. Technically it was a simple script, but many users believed the computer understood them. That reaction, arguably, created the foundation of what we see today. Humans projecting meaning and empathy onto systems that are really just rearranging symbols.

From there, progress came in leaps and bounds. In 1997, IBM's Deep Blue defeated world chess champion Garry Kasparov[6], showing that brute force computation could outmatch human expertise in structured domains. In 2012, the ImageNet breakthrough changed the game again[7]. Deep neural networks suddenly achieved unprecedented accuracy in image recognition, sparking the modern era of deep learning.

With advances in computing power in the 2000s, particularly the use of GPUs to process large datasets, the pace of AI development began to accelerate.[8] In 2017, the introduction of transformer technology changed the field again and made today's large language models possible. Just a year earlier, in 2016, DeepMind's AlphaGo had stunned the world by defeating Lee Sedol, one of the greatest players of all time.[9] The system's creative and intuition-like moves showed that AI could not only replicate human strategies but generate new ones. And by 2022, AI had moved from background automation into everyday life. Generative AI tools like ChatGPT, Midjourney, and Microsoft Copilot became mainstream, turning AI into a collaborator world-wide.

Each milestone changed how we think about machines, and each raised new questions. If computers can win at chess, generate fluent text, and create lifelike images, what does that mean for humans? We stand at the edge of understanding how AI will fit into our world and shape our future. What is clear, AI has moved from experimentation to expectation.

The AI Journey

Implementing AI is a journey. Like any journey, it requires planning, direction, and a clear understanding of the route ahead. Along the way, you will encounter terms that are often used interchangeably, even though they mean very different things.

In this chapter, I will introduce several key terms to help make sense of the full AI journey. But before we explore them, there is one in particular that can cause confusion: AI Model. Technically, the AI model is the engine, the part that powers intelligence. For example, GPT is the model that drives ChatGPT. However, in marketing, people often refer to ChatGPT as "the AI model." This use of language makes the confusion understandable.

In reality, ChatGPT is the product, not the model. It is the car built around the engine. The model (GPT) provides the intelligence and capability, while the product (ChatGPT) delivers the experience through its interface, usability, and controls. It adds safeguards such as content filters, moderation systems, and access controls that make the underlying model safe and practical for everyday use.

Throughout this book, I have maintained this distinction. I will use the technical term AI Model when referring to the underlying engine, and the term AI Car when referring to the product or application built around it.

This distinction is important, not just for clarity, but for understanding where accountability lies. Models are built and trained by vendors, but the systems and products they power are governed by the organizations that choose to deploy them.

Before we move further into terminology, it is worth remembering that implementing AI is not about the model alone. It is about the full system: the car, the engine, the driver, and the road travelled. This is the journey we are about to explore.

Algorithms, Automation & AI

The term algorithms, automation and AI is used extensively, but they are not the same thing. Not everything is AI. As Emily Bender and Alex Hanna point out in The AI Con, "AI" is often used more as a marketing term than a precise technical description.[10] When you hear the term AI, ask yourself, "What does this really mean?"

Algorithms are the heart of artificial intelligence. They keep it moving, adapting, and learning. On their own, algorithms are simply sets of instructions that tell a system how to perform a task. A spreadsheet formula that calculates payroll is an algorithm. It does the same thing every time, no matter what.

Some algorithms are very simple. Take, Linear regression. It is one of the simplest algorithms and possibly the easiest to explain.[11] Linear Regression looks for a relationship between two things and predicts an outcome based on that pattern. For example, it might be used to estimate the price of a house based on how many bedrooms it has. The algorithm draws a straight line through past data to capture the general trend. That line can be written as: $y = mx + b$.

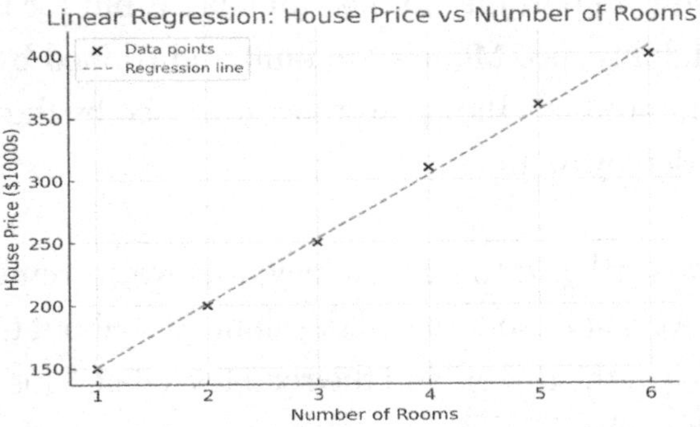

Fig 1: Linear Regression

Fig. 1 illustrates how linear regression (straight line) works in practice. If we want to estimate the price of a three-bedroom house, the algorithm looks at the slope (m) and the intercept (b). The slope shows how much each extra room adds to the price, while the intercept is the point where the line crosses the y-axis. The base represents the starting value when no rooms are counted. Other algorithms look more complex. The Gaussian distribution, often called the bell curve (curved line), is written like this:

$$f(x) = (1 / (\sigma\sqrt{2\pi}))\, e\hat{}(-(x - \mu)^2 / (2\sigma^2))$$ [12]

Linear regression and Gaussian distribution are both embedded in mathematical probability, but they serve very different purposes in practice. Linear regression is about relationships. It can estimate how much each extra bedroom adds to the price of a house, showing a direct link between variables. A Gaussian distribution is about variation. It describes how prices (or exam results, or salaries) cluster around an average, with most values in the middle and fewer at the extremes. In housing, for example, it can show the spread of property prices across Australia, from the most economical to the most expensive, and help identify what is typical versus what is outlier.

Both methods can be applied to the same domain but they answer different questions. Regression explains relationships, while distributions reveal patterns.

The distinction between the two illustrates why the choice of algorithm matters. Some algorithms are simple and interpretable, designed to provide clear cause-and-effect insights. Others are complex, uncovering hidden relationships that humans might never see. Each serves a different purpose and shapes how an AI system learns, reasons, and behaves.

David Harel states that algorithms are more than a branch of computer science.[13] It is its core. Charles Babbage described it as the "analytical engine," the mechanism that transforms logic into action.[14] There are hundreds of algorithms beyond these examples, each with strengths, limitations, and trade-offs. A business may not always choose which algorithms its vendors or partners use, but boards and executives should still understand enough to ask informed questions. Knowing whether a model is rule-based, probabilistic, or neural helps reveal how much control the organization truly has over its outcomes. Algorithmic literacy is not about writing code. It is about understanding that every algorithmic choice influences accuracy, bias, and trust.

Automation builds on the concept of algorithms. In short, automation is a repeatable set of instructions.[15] A robotic arm on a factory floor can assemble cars all day long, following a process it was programmed to repeat. It extends the power of algorithms by carrying out instructions automatically and at scale. But it is still rule-following. It never adapts on its own.

AI is different. Traditional automation is deterministic. If you give it the same input, it will always return the same output. AI is probabilistic. It blends algorithms, neural networks, and modern hardware to learn from data, adjust to patterns, and operate in environments where no two inputs are the same.[16] That is what makes AI powerful, but also unpredictable.

Let's explore how AI differs from traditional automation. Consider a document review tool. An automated system might be programmed with a catalogue of standard contract terms. It can scan a contract, sort clauses into categories, and suggest standard terms that are missing. That is useful, but it is not AI. It simply follows predefined rules. It cannot recognize unusual wording that exposes the business to risk, nor can it highlight clauses that deviate from accepted practice.

AI works differently. It can learn from thousands of contracts and their outcomes, recognize subtle variations in language, and flag clauses that are likely to be contentious. More importantly, it can interact with the user. A director could ask, "Why is this clause risky?" and the AI could explain, drawing on the patterns it has learned. That ability to adapt, reason, and respond makes AI more than automation.

When someone tells you a product is "AI-powered," don't just accept the label. Ask what it does.

- Is it calculating a value, like predicting the price of a house?
- Is it repeating the same task, like an automated process?
- Is it learning from patterns, adapting to new information, or even holding a conversation?

The answer will tell you whether you are dealing with an algorithm, automation, or true AI, and it will guide the level of oversight that is required.

AI Systems

When people talk about AI, they often focus on the model. Names like GPT-4, Gemini, or Claude are thrown around as if the model alone defines what AI can do. But organizations do not govern models. They govern systems.

An AI system is the whole experience. It is not only the model that matters. The data that goes in, the way the system is used, and the decisions that come out all shape the outcome.[17] AI systems are what customers, employees, and regulators actually experience.

Take Amazon as an example. It is not only a website. It is the full experience. Customers browse online, place an order, pay securely, and then the product is moved through warehouses and logistics.[18] The order arrives at your door and finally you open the box. Customers judge Amazon on the whole journey, not on one step.[19] The AI system works the same way. A model is one part, but the system is the full end-to-end experience, from data to decision to outcome.

Understanding the difference matters. Regulators and customers do not ask whether you chose a good model. They ask whether the full system worked as expected. If a customer is wrongly denied a loan, or a patient is mis-triaged, the defense "we picked the right model" will not hold. What matters is how that model was integrated, monitored, and governed as part of a wider system.

Fraud detection is a good example.[20] A model can be trained to spot suspicious transaction patterns, but detection alone does not prevent fraud. The outcome depends on the practices around it. If thresholds are set too low, the system generates endless false alarms. If they are set too high, genuine fraud can slip through. Once an alert is raised, the way the organization triages it can make all the difference. A clear process must ensure that the right people review the case quickly.

Handling false positives is equally important to ensure customers are not unfairly blocked or penalized. And the system is bigger than any single alert. It includes millions of transactions flowing in, dashboards for compliance teams, escalation routes to staff, and reports for regulators. What customers and regulators experience is the system as a whole, not just the model.

The same is true in healthcare. A model may recognize symptoms from patient descriptions.[21] Practices determine how reliable it is and if doctors stay involved. But the system is what matters to patients: the checks on data, the escalation when errors occur, and the record-keeping that ties back to treatment. Patients do not see the model. They 'feel' the system.

As a result, when risks are assessed, IT infrastructure or cybersecurity teams are not looking only at the AI model. They are reviewing the entire system, from the data flows, integrations, safeguards, and outputs to the vendors, cloud providers, and storage environments that support them. Their goal is to identify weaknesses that could place the business at risk.

This is why I continue to refer to these ecosystems as AI Systems rather than simply AI products or models. From a security and governance perspective, the system includes everything that connects to the AI's function and outcome. The product that the users see, the vendors and third parties involved, the data pipelines that feed into it, and the infrastructure that stores and processes information. Every one of these elements can become a source of risk or failure. A vulnerability in a vendor's API (Application Programming Interface), a misconfiguration in a storage layer, or a gap in data handling can undermine even the most capable model.

Boards and executives should view AI as an interconnected system of dependencies. Resilience depends on understanding how each part contributes to the whole environment. This includes how failure in one can cascade through to the rest. When governance focuses only on the visible product, it misses the unseen layers where security and accountability truly reside.

AI Models v AI Practices

The terms AI model and AI practices are often mistaken to be the same thing, but they are not. If the AI system is the whole car, then the model is the 'engine', the core component that provides capability. GPT-4, Claude, and Gemini are examples of these 'engines', each designed and released by their vendors. Just as Tesla, Ford, and Volkswagen produce cars with different engines, AI vendors build models with their own design philosophies, architectures, and limits. But naming the car's engine alone does not explain how it will be tuned, maintained, or driven. That is the role of AI practices.

A model provides capability. It can recognize faces in images or flag suspicious financial transactions. But capability alone is not enough. What matters is how that capability is applied, governed, and monitored. Clearview AI shows the distinction. Its model can match faces from a database of images, yet the same model would not be appropriate for reading medical scans.[22] Different contexts require different training data, safeguards, and oversight.

The Robodebt scheme in Australia shows what happens when practices fail completely. The model calculated debts, but without safeguards and oversight the result was widespread harm, legal challenges, and ultimately a Royal Commission.[24] A label does not tell the full story. An AI system is like a car. It is the whole 'journey'. It is the experience from body to engine to the way it is driven. The model is the engine inside, built to deliver power, but an engine alone does not determine safety or performance. AI practices are the equivalent the parts, the tuning, maintenance, and how the car is driven on the road. Two cars may carry the same logo, but how they are maintained and operated makes all the difference. The same is true for AI. An AI model is not an AI practice, and both must be evaluated to understand the real risks and responsibilities.

The name of a model does not tell the whole story. A Tesla's engine is tuned for efficiency and electric speed, but depends on charging. A Ranger's engine is tuned for torque and rough terrain, but less efficient in cities. In the same way, GPT-4, Claude, and Gemini differ in architecture, training data, and intended use cases. One model may be better at reasoning across long documents, another in generating content, and another more versatile in coding. The choice between each depends on context such as efficiency, versatility, or raw power. The difference between models themselves can be linked to the approach to AI Practices.

AI practices are the methods used to train, fine-tune, deploy, and govern a model.[23] They shape how the model behaves. Two organizations may both say they use GPT-4, but that can mean very different things. Through ChatGPT, the model comes pre-packaged. What you see is what you get. In Microsoft Copilot, GPT-4 is embedded into productivity tools. And in an enterprise setting, the same model might be pulled in through an API, fine-tuned with company data, paired with human review, and actively monitored. The model name may be the same, but the purpose, context, and strategy for its use can be very different. The difference is the shift in risks, safeguards, and governance responsibilities.

References

1. Stuart Russell and Peter Norvig, *Artificial Intelligence: A Modern Approach* (4th ed, Pearson, 2023).
2. Keyword Results for Artificial Intelligence' (WordStream, n.d.) https://tools.wordstream.com/fkt?website=artificial+intelligence&cid=&camplink=&campname=&geoflow=0.
3. A M Turing, Computing Machinery and Intelligence' (1950) 59 *Mind* 433 https://courses.cs.umbc.edu/471/papers/turing.pdf.
4. John McCarthy et al, A Proposal for the Dartmouth Summer Research Project on Artificial Intelligence (1956) https://home.dartmouth.edu/about/artificial-intelligence-ai-coined-dartmouth.
5. Daniel Jurafsky and James H Martin, *Speech and Language Processing* (2nd ed, Dorling Kindersley (India) Pvt Ltd, 2023).
6. Deep Blue Defeats Chess Champion' Chess.com https://www.chess.com/terms/deep-blue-chess-computer.
7. ImageNet' Pinecone Learning Series https://www.pinecone.io/learn/series/image-search/imagenet/.
8. Chris Miller, *Chip War: The Fight for the World's Most Critical Technology* (Simon & Schuster, 2022).
9. James Purtill, AlphaGo Marked the Birth of Modern AI: This Is the Moment the World Changed' (ABC News (online), 25 October 2023) https://www.abc.net.au/news/science/2023-10-25/alphago-versus-lee-sedol-when-everything-changed-for-ai/102988050.
10. Emily M Bender, *The AI Con* (The Gradient, 2021).
11. Richard Szeliski, *Computer Vision: Algorithms and Applications* (Springer, 2022).
12. Ibid.
13. David Harel, Algorithmics: The Spirit of Computing (Addison Wesley Longmand Ltd, 199).Ibid.
14. Bernd Ruecker, *Practical Process Automation* (O'Reilly Media, 2021).
15. Stuart Russell and Peter Norvig, *Artificial Intelligence: A Modern Approach* (4th ed, Pearson, 2023).
16. Organisation for Economic Co-operation and Development (OECD), Updates to the OECD's Definition of an AI System Explained' (OECD AI Policy Observatory, 2023) https://oecd.ai/en/wonk/ai-system-definition-update.
17. Dana Mattioli, *The Everything War: Amazon's Ruthless Quest to Own the World and Remake Corporate Power* (Little, Brown and Company, 2024).

18. Mark Wilkins, *Learning Amazon Web Services (AWS): A Hands-On Guide to the Fundamentals of Aws Cloud* (Pearson Education, 2020).
19. Aisha Abdallah et al, Fraud Detection System: A Survey' (2016) 68 *Journal of Network and Computer Applications* 90 https://www.sciencedirect.com/science/article/abs/pii/S1084804516300571.
20. Junaid Bajwa et al, Artificial Intelligence in Healthcare: Transforming the Practice of Medicine' (2021) 8(2) *Future Healthcare Journal* e188 https://pmc.ncbi.nlm.nih.gov/articles/PMC8285156/.
21. Ian Goodfellow, Yoshua Bengio and Aaron Courville, *Deep Learning* (MIT Press, 2016).
22. Royal Commission, Robodebt Scheme (2023) https://robodebt.royalcommission.gov.au/publications/report.
23. Clearview AI' (Official Website) https://www.clearview.ai/.

*"Models drive the logic.
Products define the experience."*

CHAPTER 2

Types of AI Products

Artificial intelligence looks simple when it appears as a chatbot on a screen or as a friendly assistant inside a productivity tool. To the everyday user, it feels like magic. Type a question and receive a fluent response. But what you see on the surface, such as with ChatGPT, Copilot, or Gemini, is only the vehicle's exterior. Underneath lies the engine. As discussed previously, this is the model, designed and assembled to recognize patterns, generate responses, and perform tasks. Understanding this difference is important because boards and executives often see only the polished car on display, not the engine inside that determines its power, reliability, cost and safety.

To understand this better, let's return to the car analogy. A Volkswagen Golf and a Tesla Model S are both cars, but their mechanics and design philosophies are radically different. One is tuned for efficiency and everyday use; the other for speed and innovation in electric driving. Likewise, AI models vary. Some are optimized for creativity, others for reliability, others for speed. A car's name tells you something, but its true performance depends on what is under the hood and how it is driven.

While cars are different, they share the same engineering principles. They rely on engines, power systems, and frameworks that evolved over decades of innovation. AI models are much the same. Beneath their branding, they are built on a common foundations of architecture and methods that define how they process information, learn from data, and perform tasks.

Modern AI models have a shared foundation. In 2017, Google researchers published *"Attention Is All You Need,"* [1] introducing the transformer architecture. [2] This novel approach relies solely on attention mechanisms that process entire sequences in parallel instead of step-by-step, as older models did. This design allowed AI systems to capture long-range relationships in data, such as connections between words spread across a sentence or document, while dramatically increasing efficiency and scalability.

The transformer's innovation lies in its self-attention mechanism, which enables the model to weigh the importance of each data element in context. Whenever someone adds a prompt or document, the transformer gives each word a value of importance, allowing the model to focus on the most relevant words for understanding or generating responses. This capability allows the model to train on enormous datasets with superior accuracy and speed compared to previous models like RNNs (Recurrent Neural Networks) or CNNs (Convolutional neural network). I will discuss CNN's later in this book.[3]

Transformers quickly became the foundation for today's leading language models, including GPT, Gemini, Claude, and Perplexity. These models build on principles set out in the original paper. While Google introduced the technical architecture, OpenAI's ChatGPT brought transformers to widespread public attention. This shows how broad adoption often follows popularity, rather than a careful balance of strategy and risk.

AI models have been built and released in the middle of a high-stakes race for dominance.[4] Vendors make choices about training data, architecture, and deployment strategies that reflect their commercial goals. These choices can ripple downstream to any organization that adopts the model. A model trained primarily on English-language data may stumble in multilingual contexts. One optimized for speed may be more error-prone.[5] Another marketed as "safe" may come with constraints that limit creativity.

Because these design decisions are made upstream, every organization that adopts a model inherits part of the vendor's intent, priorities, and risks. Selecting an AI model is therefore not just a technical decision. It is a governance issue with consequences for cost, security, control, and long-term alignment with strategy. Adopting a model is entering a relationship with its vendor — one defined by contracts, infrastructure, and incentives.

Chapter 2 introduces the major "cars" on today's AI track. Each can be thought of as a different car in the race: the disruptor, the integrator, the multimodal challenger, the safety-first counterweight, and the transparent outsider. None is neutral. Each reflects the priorities of the vendor behind it. The goal is not to turn board members and executives into engineers but to equip them with the literacy to see past the brand names, look under the hood, and ask the questions that matter.

AI Products currently in the Market

Before diving in, it is worth clearing up a common misconception. What most people call an AI model in practice is usually an AI product, or arguably an AI System. It is the full car, not just the engine. ChatGPT, Copilot, Gemini, Claude, and Perplexity are all systems built on top of underlying models such as GPT-4 or Claude Opus.

The AI model is the engine. It provides the intelligence, the capability, and the power. The AI system is the car. It is the full experience delivered to the user, from the interface and data flows to the outcomes, safeguards, and risks. The user doesn't see the engine. They experience how it performs.

Throughout this chapter, we will treat each of these "cars" as examples of how different vendors have packaged their engines into products. Our aim is to move beyond branding and analyze how design decisions in data, architecture, and deployment affect risk, performance, and trust.

The Disruptor - ChatGPT

When ChatGPT appeared in late 2022, it seemed less like the release of a product and more like the ignition of a cultural moment.[6] For the first time, millions of people could interact with an AI system that felt conversational, fluent, and accessible. The idea of "artificial intelligence" had hovered in public imagination for decades, but always as something distant.

Until 2022, AI was nothing but a lab experiment, a futuristic promise, or a Hollywood movie. Suddenly it was within everyone's reach. You didn't need to understand coding, machine learning, or neural networks. You just needed to type.

The speed of adoption was staggering,[7] within weeks, ChatGPT became a household name. University students were experimenting with essays, knowledge workers were testing whether it could draft reports or summarize documents, journalists were using it for background research, and parents were asking it to write bedtime stories.

The effect was similar to the release of the first web browser in the 1990s.[8] At first it was an obscure technology. Type something in a box and suddenly one was presented with options.

I have seen firsthand how quickly AI has been adopted and the concerns that followed. At university, students were using ChatGPT to write assignments and even assist during online exams. This raised immediate questions about integrity, oversight, and responsible use. The same challenge now faces boards, as they are challenged with how to govern against misuse within their own organizations.

The appearance ChatGPT as an overnight success, is misleading. It was definitely not. It was the result of years of research, iteration, and scaling by OpenAI.[9] The Generative Pre-trained Transformer, or GPT, had already evolved through several versions before it was wrapped in a conversational interface and released to the public.[10]

ChatGPT is not a single model but a system built around GPT. The "chat" interface provides accessibility, while the GPT model performs the language generation behind the scenes. What changed was not the model itself, but the accessibility. By adding a simple conversational layer, OpenAI turned a research model into a product anyone could use.

The complexity of billions of parameters and vast training data was hidden behind something familiar: a text box. To users, it felt like magic and marked the moment AI moved from a technical tool to a strategic force.

ChatGPT performs efficiently, but its operation is far from economical. Every prompt, whether a single line or a full essay, triggers massive computations across Microsoft's Azure cloud. Tens of thousands of

GPUs run in parallel, consuming enormous amounts of electricity and driving costs into the hundreds of thousands of dollars per day.[11]

Microsoft's multi-billion-dollar investment in OpenAI, spread across equity, infrastructure, and credits, allowed those costs to be subsidized in the short term.[12] Consumer success stories in AI rest on extraordinary infrastructure commitments. These costs do not disappear. They resurface in enterprise licensing, subscription fees, and usage charges. Running ChatGPT is like driving a supercar. With every dazzling performance, burns fuel at a rate few can afford.

This adoption presents challenges for organizations that seek to harness AI responsibly. While the technology offers remarkable capability, it also brings financial, ethical, and operational trade-offs that cannot be ignored.

The governance lesson is twofold. First, AI adoption is never free.[13] A simple website sits on vast infrastructure, and those costs eventually flow to customers. Second, consumer virality is not the same as enterprise reliability. A system that amazes students with essays may raise serious concerns in busineses such as law firm, financial institutions or even online retail. Directors must look past the hype and ask: How is it trained? What does it cost to run at scale? What risks arise if staff use it informally without oversight?

ChatGPT changed the narrative about AI from 'myth' to practice. It forced organizations to reckon with the fact that the technology was no longer experimental, but commercial and widely available. Like the sports car that captured everyone's attention, it dazzled with speed and accessibility, putting AI in the hands of millions almost overnight. Yet, like any high-performance vehicle, it came with high running costs, heavy energy demands, and limited practicality for enterprise use.

Without proper training, handing it to unprepared users was like giving a sports car to a learner driver and expecting no accidents.

The Integrator - CoPilot

If ChatGPT was the disruptor, then Microsoft's Copilot would be considerd as the enterprise integrator. Microsoft recognized that organizations would need the same generative power but in a way that respected corporate boundaries. [14] A public chatbot may be intriguing, but businesses know they can not rely on tools where employees uploaded sensitive contracts or internal reports into a website outside of the company's control. Microsoft's answer was Copilot. It uses the same underlying GPT models, but embedded directly into the Microsoft 365 suite.[15]

In Word, Copilot drafts and edits text. [16] In Outlook, it summarizes email threads and proposes responses. In PowerPoint, it builds presentations from a simple prompt. By bringing the model into the applications used by the business, Microsoft lowered adoption barriers. Additionally, they gave assurance that data would remain within the company's Microsoft 365 tenant and processed under Azure's compliance and governance frameworks.

Microsoft positioned CoPilot as both as a feature and a pricing lever. Its integration into Microsoft 365 ensures that AI is not only sold as a tool but as a way to structure licensing revenue.[17] As such this turns AI adoption into a financial and strategic decision, not just a technical one.

If ChatGPT is the sports car that dazzled the public, Copilot is the four-wheel-drive utility vehicle. It is built for enterprise terrain. A robust, practical, and designed vehicle to integrate with heavy workloads. Like any 4x4, it comes with high fuel costs and needs more upkeep. It is excellent under the right conditions, the question is whether efficiency trade-offs are acceptable.

Microsoft did not stop at the Microsoft 365 environment. It extended the Copilot brand into specialized products. One of the most influential was GitHub Copilot, designed for software developers. Microsoft integrated Copilot directly into tools like Visual Studio Code.[18] It could autocomplete code, suggest improvements, and accelerate delivery. However, it raises questions in relation to vulnerabilities hidden in code or open-source components which had unclear licensing. Questions are also raised as to whether developers could copy sensitive company code into a system they did not fully control. The productivity gains were significant, but so were the exposures in intellectual property, compliance, and security.

Another Microsoft product variant is Copilot WebChat which is accessible through Bing Chat or directly via a browser such as Edge, Chrome, or Firefox.[19] Unlike GitHub Copilot or the Microsoft 365 suite, WebChat is placed in a gray zone. In some organizations it is included as part of a formal rollout, while in others, staff can access it independently to test AI capability without waiting for enterprise integration. The downside is governance. There is a risk that employees could share sensitive data without oversight, creating significant compliance and confidentiality issues. For many industries, this lack of oversight represents a significant governance concern.

Together, the three products illustrate a key point. The same 'engine' can create very different risk profiles depending on how it is packaged into systems and deployed. A report that says "we're using Microsoft Copilot" tells only half the story. However, boards should ask more pointed questions like:

- Is the organization investing in enterprise-grade Microsoft 365 Copilot with data protection and compliance built in?
- Is it allowing developers to use GitHub Copilot with intellectual property risks?
- Or is it letting staff experiment with Webchat, effectively stepping outside formal controls?

The lesson is not to avoid these tools but to understand the terrain.

Microsoft's strategy across this family of products also reveals the deeper dynamic of the AI race. By embedding Copilot into Microsoft 365, GitHub, and even the browser, Microsoft has created an ecosystem that enterprises find difficult to exit. This is not only about productivity, it is also about locked-in agreements. Once a company adopts Copilot across multiple fronts, switching vendors could become disruptive and costly. The strategic question for boards is whether dependence on a single vendor is acceptable. The convenience of integration should be balanced against the loss of flexibility in the future.

As stated, copilot is the four-wheel-drive of enterprise AI. It is built for tough terrain, capable of carrying heavy workloads and navigating complex environments by embedding AI directly into the tools businesses already use. Like any 4x4, it is powerful and dependable, but it comes at a cost. It consumes significant resources[20], requires ongoing maintenance, and once it becomes part of daily operations, changing vehicles is no easy task.

The Multimodal Challenger - Google Gemini

If OpenAI and Microsoft captured the spotlight with ChatGPT & Copilot, Google's story is a little more paradoxical. On the one hand, no other company has contributed more to the foundations of modern AI than

Google.[21] On the other, Google could be portrayed as having hesitated at a crucial moment, while OpenAI raced ahead with products built directly on research Google itself had pioneered.

While we cannot know the internal decisions, it is reasonable to consider that Google applied a more cautious strategy, perhaps shaped by lessons from the past such as browser wars, privacy debates, and antitrust scrutiny.[22] Rather than rushing to be a trend-setter, Google may have chosen to act as a fast follower, waiting to see how the market and regulators responded before scaling.[23] Google's actions provides a lesson. Moving slowly is not always weakness, sometimes it is strategy supported by a calculated balance of risk, timing, and sustainability.

Gemini represents Google's effort to change that. Unlike earlier "cars' it was built from the ground up with multimodality in mind, meaning it can understand and generate more than one type of input and output. In practice, this allows Gemini to process text, images, and code, and in some versions, audio and video. Packaged as the Gemini system, it enables users to summarize reports, interpret screenshots, analyze spreadsheets, and explain images, all within the same conversation.[24]

Just as Microsoft tied Copilot into Word and Excel, Google has embedded Gemini into its Workspace (Gmail, Docs, Sheets, Meet, Drive, Calendar, Chat, and Slides).[25] Employees can draft emails, summarize threads, generate reports, or analyze data without leaving their familiar tools. The appeal is obvious. Using Gemini does not require retraining staff on new platforms. It creeps quietly into current workflows, promising productivity gains without disruption.

Integration comes with trade-offs. The more embedded a model is into core workflows, the harder it is to exit later. Adopting Gemini means aligning closely with Google's ecosystem, its data governance standards, and its infrastructure. For companies that are already reliant on Google Workspace, this may feel seamless. For others, it raises the risk of vendor lock-in. Switching costs are not only financial but operational. Once staff adapt to AI workflows, reversing the implementation is not just costly, it is disruptive.

While Gemini's power lies in its integration within Google Workspace, the integration should gives some boards caution to pause. Google is a pioneer in AI, yet it carries a complicated history on data use. Historically, Google scanned user emails to deliver targeted advertising.[26] Google Maps faced scrutiny for collecting location data without sufficient consent.[27] The company's success has been built on turning data into revenue, sometimes at the expense of transparency. For organizations now considering embedding Gemini into their most sensitive workflows, cannot be ignored that legacy. Trust is not only about technical performance, it is about governance, accountability, and a vendor's track record.

Another defining feature of the Gemini system is its real-time grounding in Google Search.[28] While the underlying model provides reasoning and generation, the system augments this with live retrieval from the web. This is a double-edged sword. On one hand, it makes Gemini highly valuable for monitoring markets, analyzing trends, or tracking competitors. On the other hand, it introduces dependencies and risks. If the AI model grounds its answers in web data, what assurances is there in relation to accuracy, origns, or security? What controls exist to prevent misinformation being repeated as fact?

Pricing follows a two-tier model. For individuals, Gemini comes through Workspace subscriptions, usually at extra cost.[29] For organizaitons,

Google offers Vertex AI, a platform for fine-tuning models, building pipelines, and scaling use.[30] Charges are usage-based, similar to cloud services, which makes Gemini flexible but can also make budgeting tricky. Without clear policies, costs can escalate quickly.

Copilot and Gemini illustrate clearly two different strategies. Copilot is integration-first. It takes an external engine (OpenAI's GPT) and embeds it into the tools employees already use, wrapping governance and compliance around it. Gemini is foundation-first. It uses Google's own multimodal engine as the core, then packages it into systems delivered through Workspace or Vertex AI. One approach prioritizes seamless embedding, while the other emphasizes building on a powerful foundation and extending outward. The decision is less about brand dominance and more about alignment. Ultimately, it comes down to choosing the ecosystem that best matches the organization's long term strategy, cost discipline, and risk appetite.

Gemini is like the family car. It blends versatility with everyday practicality, handling different formats such as text, images, and code within the same ride. For organizations already in Google's ecosystem, it slips seamlessly into daily use. But once inside Google's garage, leaving can be disruptive and costly. Convenience comes with commitment.

The Safety-First Counterweight - Claude

If Copilot is the enterprise embedder and Gemini the multimodal challenger, Claude represent the safety-first counterweight. Where others raced ahead to capture markets, Anthropic built its strategy on restraint, caution, and governance by design.

Claude's developers asked a more philosophical question, "How do we build an AI that is not only powerful but aligned with human values?"

This shift in emphasis is important. It highlights that not all AI models are created with the same goal. The trade-offs between capability, safety, and trustworthiness can offer a distinctive value.

Anthropic was born out of a fracture within OpenAI.[31] In 2021, a group of senior researchers, including Dario Amodei, Daniela Amodei, and Tom Brown, left to found their own company. [32] Their concern was that OpenAI was moving too quickly, scaling larger models without sufficient attention to alignment and control. Anthropic's mission became to build systems that were more predictable, steerable, and transparent. Their AI Model would not only generate impressive results but do so in ways that reduced the risk of harmful or misleading outputs.

Claude, named after Claude Shannon, the father of information theory, reflects that philosophy.[33] Unlike ChatGPT, which was fine-tuned through reinforcement learning from human feedback,[34] Claude was trained using "constitutional AI." Rather than relying solely on human trainers to adjust behavior, Anthropic provides the model with a written set of principles, a kind of constitution, and teaches it to apply them consistently.[35] The result is a model designed to produce fewer harmful responses, explain its reasoning more clearly, and behave more transparently.

In practice, Claude has developed a reputation for a more measured and thoughtful tone. Its latest version, Claude 3 Opus, rivals GPT-4 on many benchmarks while offering longer context windows. It can hold and analyze more information within a single conversation, with a deliberate emphasis on aligning responses to user intent.

Claude's safety-first approach is not without limits.[36] Like any large language model, it can still misinterpret prompts or generate inaccurate outputs. In 2024, researchers uncovered vulnerabilities affecting several

Large Language Models (LLMs), including Claude, where malicious prompts could trigger data leakage.[37] Anthropic responded quickly, but the episode underscored an uncomfortable truth. Even safety-focused models remain exposed to risk.

The next year, the company also agreed to pay millions to writers who claimed that their work had been used in training without consent.[38] This move, arguably, sets it apart from larger rivals, who have often fought such claims in court. Together, this reflects the tension of Anthropic's position. It strives to embody responsibility while competing in a marketplace that rewards speed and scale.

Anthropic's commercial approach also reflects its safety-first philosophy. Claude is accessible in tiers.[39] One can choose from a free web interface for light use, a Pro subscription for heavier users, and enterprise APIs with compliance features and access controls. Unlike OpenAI, which is bound to Microsoft's Azure, Anthropic has aligned itself with Amazon, securing billions in investment and making AWS its primary cloud partner. [40] This type of independence gives boards an alternative to a Microsoft-Google duopoly. However, it still introduces reliance on Amazon's infrastructure..

What sets Claude apart is less its technical specifications than its strategic posture. Anthropic has positioned itself as a counterweight in the AI race. Its emphasis is on alignment, safety, and transparency by offering organizations a different philosophy to consider capability balanced with governance. Whether that emphasis consistently holds up in practice is still an open question, but the positioning is commendable. This demonstrates that the AI race is not only about building ever larger AI models or capturing market share quickly. It is also about the competing values and governance that underpin these systems.

Claude's story highlights three key points. First, AI is not a monolith. It reflect the values, incentives, and philosophies of their makers. An AI model built by a company chasing consumer scale will behave differently than one built by researchers prioritizing safety. Second, "responsibility" is itself a competitive differentiator. Just as companies compete on speed, features, and cost, they now compete on trust. Boards must recognize this and decide whether their own organizations value raw capability, guardrails, or a balance of both. Third, no vendor can remove governance responsibility from the board. Even if Claude reduces certain risks, oversight, monitoring, and accountability still rest with the organization.

Claude represents a philosophical fork in the road. One path is with bigger models with greater capability but unknown risks. The other is cautious, emphasizing alignment, safety, and governance. Both carry costs. What matters for boards and directors, is recognizing that there is a choice. The decision is not only about what an AI model can do, but about what values the organization wants reflected in the technology it adopts.

Claude is the Volvo of AI. It was built with alignment, safety, and governance at the center of its design. It does not promise the fastest acceleration but emphasizes protection, predictability, and trust. For organizations that value reliability and guardrails, Claude offers a cautious but steady ride.

The Transparent Outsider - Perplexity

While we've discussed giants like OpenAI, Microsoft, and Google, there is a quieter player that has carved out a unique position. Perplexity is neither the biggest nor the best-funded, but it has won a following for one simple reason. It shows its work. Unlike Claude, which emphasizes alignment and safety, Perplexity emphasizes openness. Anyone can try it immediately without enterprise contracts. For users, that combination of accessibility and accountability has made it a trusted companion. Why? Because it does not demand personal information to function and is transparent in its answers.

Where ChatGPT, Copilot, and Gemini generate answers with confidence, Perplexity shows where the answers come from. In organizations such finance, law, and healthcare, directors ahould ask: What is the source and can we verify it? Perplexity enables that discipline but it is not infallible. Its sources can still be unreliable, but it shifts the relationship from blind trust to verification.

Perplexity was founded in 2022 by Aravind Srinivas, a former researcher at OpenAI and DeepMind.[41] Together with his colleagues from Meta and Quora, the vision was not to build the largest model but to build the most useful one. Perplexity created what is known as retrieval-augmented generation (RAG)[42] by combining open-weight AI models like Meta's LLaMA with its own retrieval systems. The idea is to supplement the model's static training with real-time access to web content, so the answers are not limited by a cutoff date and are anchored in current information.[43] For busy specialists, such as analysts, researchers, executives, this combination was powerful. The system felt less like a chatbot making things up and more like a conversational search engine.

Perplexity highlights a subtle but important fact. Smaller players can shape the governance conversation as much as larger ones. While OpenAI was dazzling the public with scale, Google was integrating Gemini into Workspace, Perplexity was quietly winning loyalty by focusing on both transparency and trust. In some industries, that trust factor is decisive. For example, a law firm may prefer a slightly less powerful AI model that provides citations over a AI model that does not.

The company's business model follows similarities to other AI companies. Perplexity offers a freemium service for casual use, a Pro subscription for individuals, and enterprise plans with API access for organizations. At first glance, its pricing is straightforward and competitive.[44] It is cheaper than enterprise licenses for Copilot or Gemini. But costs can still escalate quickly in enterprise deployments, especially if Perplexity is embedded into customer facing service platforms or research workflows where usage is heavy. Like all AI services, the economics are usage-based. The more prompts, the more tokens processed, the higher the bill.

Perplexity's approach carries its own risks. Using live web content, the model inherits the vulnerabilities of the web itself. Sources are not always accurate. Some may be bias, outdated, or deliberately misleading. Boards must understand that transparency is not the same as reliability. A wrong citation, left unchecked, can be just as damaging as no citation at all. The governance challenge is to ensure that staff do not mistake the presence of a citation for proof of truth. Verification still requires judgment.

Perplexity also raises pointed questions of data governance. Prompts entered into the system are processed on its servers, which means that organizations face the same concerns they would with any AI service. Where is the data stored, under what jurisdiction, and for how long? According to Perplexity's data policy, user interactions may be collected and used to improve model performance unless the user is on a paid Pro or Enterprise plan, where stricter privacy controls apply.[45] This distinction highlights that not all usage is treated equally. Even if a user deletes a chat, it may be retained temporarily for operational or diagnostic purposes. Oversight therefore remains critical. Can administrators audit staff usage, enforce deletion policies, or prevent employees from accessing Perplexity outside the corporate environment?

Perplexity embodies both opportunity and caution. It proves that not all innovation comes from giants. Smaller, more agile companies can deliver models that meet specific governance needs, such as traceability and verifiability. Yet its reliance on live data and position outside the larger vendor ecosystems create different exposures. Transparency in outputs does not necessarily mean transparency in operations. Organizations must weigh the benefits of cited sources against the risks of relying on a system that draws from the open web. Cost management is another factor. Perplexity's pricing is clear on paper, but usage can escalate quickly when integrated into enterprise workflows.

It is easy to believe that only OpenAI, Microsoft, or Google matter, but smaller players are proving otherwise. For organizations whose focus is trust, and verification, AI models like Perplexity show that alternatives exist. They may not offer the same scale, but they can align more closely with organizational values.

References

1. Ashish Vaswani et al, Attention Is All You Need (2017) 31st Conference on Neural Information Processing Systems (NIPS).
2. Oswald Campesato, Artificial Intelligence, Machine and Deep Learning (Mercury Learning and Information LLC, 2020).
3. Alammar J, The Illustrated Transformer (Blog post, 26 June 2018) https://jalammar.github.io/illustrated-transformer/.
4. Robert A Manning, 'AI Race: The Promise and Perils of Techno-Utopians' (Stimson Center, 2025) https://www.stimson.org/2025/ai-race-the-promise-and-perils-of-techno-utopians/.
5. Stuart J Russell and Peter Norvig, Artificial Intelligence: A Modern Approach (Pearson, 4th ed, 2020).
6. OpenAI, 'Introducing ChatGPT' (Web Page, 2022) https://openai.com.
7. Josh Howarth, '50 New Artificial Intelligence Statistics' (Exploding Topics, 2025) https://explodingtopics.com/blog/ai-statistics.
8. Tim Berners-Lee, Weaving the Web (Harper San Francisco, 1999).
9. Bernard Marr, 'A Short History of ChatGPT: How We Got to Where We Are Today' (Forbes, 19 May 2023) https://www.forbes.com/sites/bernardmarr/2023/05/19/a-short-history-of-chatgpt-how-we-got-to-where-we-are-today/.
10. Donna McGeorge, The ChatGPT Revolution: Get Curious, Get Productive and Get Creative with AI (John Wiley & Sons Australia, Ltd, 2024).
11. Rionna Agomuoh, 'You Won't Believe How Much ChatGPT Costs to Operate' (Digital Trends, 2023) https://www.digitaltrends.com/computing/chatgpt-cost-to-operate/.
12. Microsoft, 'Microsoft and OpenAI Evolve Partnership to Drive the Next Phase of AI' (Blog Post, 21 January 2025) https://blogs.microsoft.com/blog/2025/01/21/microsoft-and-openai-evolve-partnership-to-drive-the-next-phase-of-ai.
13. OpenAI, 'ChatGPT Pricing' (Web Page, 2025) https://chatgpt.com/pricing.
14. Microsoft, 'Introducing Microsoft 365 Copilot: Your Copilot for Work' (Blog Post, 16 March 2023) https://blogs.microsoft.com/blog/2023/03/16/introducing-microsoft-365-copilot-your-copilot-for-work/.
15. Ibid.
16. Microsoft, Microsoft 365 Copilot Documentation (Microsoft Learn, 2023) https://learn.microsoft.com/en-us/microsoft-365-copilot/microsoft-365-copilot-overview.
17. Microsoft, 'Microsoft 365 Plans and Pricing' (Web Page, 2025) https://www.mic-

rosoft.com/en-au/microsoft-365/business/microsoft-365-plans-and-pricing.
18. GitHub, 'GitHub Copilot' (Web Page, 2025) https://github.com/features/copilot.
19. Microsoft, 'Microsoft Copilot for Individuals' (Web Page, 2025) https://www.microsoft.com/en-au/microsoft-copilot/for-individuals?form=MY02P9.
20. GitHub Community, 'GitHub Copilot High Resource Usage Bug Report #56755' (Online community discussion, 30 May 2023–21 April 2025) https://github.com/orgs/community/discussions/56755.
21. Eya Laouini, 'Perspective – From Deep Learning to Foundation Models' (Medium, 2025) https://medium.com/@eyai/the-rise-of-generative-ai-googles-perspective-from-deep-learning-to-foundation-models-e8078c682510.
22. Eric Schmidt and Jonathan Rosenberg, How Google Works (John Murray Publishing, 2017).
23. Alistair Barr, 'ChatGPT Is Crushing Google in the AI Race – Unless You Look at the Data Differently' (Business Insider, 2025) https://www.businessinsider.com/chatgpt-crushing-google-ai-unless-you-look-data-differently-2025-4.
24. Sundar Pichai and Demis Hassabis, 'Introducing Gemini: Our Largest and Most Capable AI Model' (Google Blog, 2023) https://blog.google/technology/ai/google-gemini-ai/.
25. Ibid.
26. ABC News, 'Google Amends Terms of Service to Include Email Scanning' (15 April 2014) https://www.abc.net.au/news/2014-04-15/google-amends-terms-of-service-to-include-email-scanning/5390454.
27. Australian Competition and Consumer Commission, 'Google Allegedly Misled Consumers on Collection and Use of Location Data' (Media Release, 2023) https://www.accc.gov.au/media-release/google-allegedly-misled-consumers-on-collection-and-use-of-location-data.
28. Google Cloud, 'Multimodal AI Use Cases' (Web Page, 2025) https://cloud.google.com/use-cases/multimodal-ai?hl=e.
29. Google, 'Gemini Subscriptions' (Web Page, 2025) https://gemini.google/subscriptions.
30. Google Cloud, 'Vertex AI Overview' (Web Page, 2025) https://cloud.google.com/vertex-ai?hl=en.
31. Anthropic, 'Introducing Claude' (News Post, 2023) https://www.anthropic.com/news/introducing-claude.
32. Samantha Stokes and Riddhi Kanetkar, 'Meet the OpenAI Mafia: These 23 Former Employees Left the AI Juggernaut and Founded Startups That Have Collectively Raised Billions' (Business Insider, 2024) https://www.businessinsider.com/former-openai-employees-who-left-launch-vc-backed-ai-startups-2024-1.

33. Time, 'Definition of Claude AI' (2025) https://time.com/collections/the-ai-dictionary-from-allbusiness-com/7273938/definition-of-claude-ai.
34. Ryan O'Connor, 'How ChatGPT Actually Works' (AssemblyAI Blog, 2022) https://www.assemblyai.com/blog/how-chatgpt-actually-work.
35. Yuntao Bai et al, 'Constitutional AI: Harmlessness from AI Feedback' (2022) arXiv preprint arXiv:2212.08073 https://arxiv.org/abs/2212.08073.
36. Time, 'Definition of Claude AI' (2025) https://time.com/collections/the-ai-dictionary-from-allbusiness-com/7273938/definition-of-claude-ai/.
37. Ravie Lakshmanan, 'Researchers Uncover Prompt Injection Vulnerabilities in DeepSeek and Claude AI' (The Hacker News, 2024) https://thehackernews.com/2024/12/researchers-uncover-prompt-injection.html.
38. Chloe Veltman, 'Anthropic Settles with Authors in First-of-Its-Kind AI Copyright Infringement Lawsuit' (NPR, 5 September 2025) https://www.npr.org/2025/09/05/nx-s1-5529404/anthropic-settlement-authors-copyright-ai.
39. Anthropic, 'Claude Pricing' (Web Page, 2025) https://claude.com/pricing.
40. Amazon Web Services, 'Anthropic's Claude in Amazon Bedrock' (Web Page, 2025) https://aws.amazon.com/bedrock/anthropic/.
41. Wendy Zeng, 'Perplexity Decoded: Lessons from the Founder-CEO and His PhD Advisor' (Berkeley SCET, 2025) https://scet.berkeley.edu/aravind-srinivas-lessons-from-building-perplexity-ai/.
42. Patrick Lewis et al, 'Retrieval-Augmented Generation for Knowledge-Intensive NLP Tasks' (2020) arXiv preprint arXiv:2005.11401 https://arxiv.org/abs/2005.11401.
43. Perplexity, 'Introducing the Perplexity Search API' (Blog Post, 2025) https://www.perplexity.ai/hub/blog/introducing-the-perplexity-search-api.
44. Perplexity, 'Enterprise Pricing' (Web Page, 2025) https://www.perplexity.ai/enterprise/pricing.
45. Perplexity AI, 'Data Collection at Perplexity' (Help Centre, 2024) https://www.perplexity.ai/help-center/en/articles/11564572-data-collection-at-perplexity.

"AI is not only one
Machine.
It is many engines
with many parts!"

CHAPTER 3

AI Practices?

AI practices should not be confused with AI models. If the model is the engine then practices are the methods and techniques that determines how it performs. Two vehicles may have the same logo, but each have differences in their pistons, fuel injectors, and steering systems can make them drive very differently. The same is true for AI. The product may be the car, but the practices are the hidden methods that shape how it actually behaves.

AI practices are the real-world methods and techniques that give an AI system its capabilities. Machine learning, natural language processing, and computer vision are examples of AI practices.[1] These are not models in themselves but the disciplines that allow models to learn, interpret, and act. They define how the "car drives," how reliable it is, and what it can realistically achieve once deployed.

By contrast, a model is the engine that powers the car. It is trained using those practices, tuned for a purpose, and released onto the road. Whether the model is built for chat, translation, prediction, object detection or image creation, every model depends on three building blocks: data, algorithms, and hardware.[2]

Data is the fuel and AI learns from the data it is given, may it be sales records, medical images, or internet text. The quality of that fuel determines how well the system performs. Algorithms are the mechanical parts, the mathematical rules that process the data and generate output [3], the equivalent of pistons, gears, and suspension.Hardware is the chassis, without GPUs, TPUs, memory, and storage, AI systems cannot operate at scale. These components provide the horsepower to run enormous datasets and calculations in real time.

It may not always be necessary to look under the hood, but its important to ask how these parts fit together. An organization that buys the same model as its competitor may face completely different outcomes depending on how the practices are applied, how data is sourced, how algorithms are tuned, and how hardware capacity is managed.

With that in mind, let us turn to the most widely used pratice that supports nearly every AI system: machine learning.

Machine Learning

Machine learning is the practice that makes artificial intelligence practical.[4] It uses algorithms to find patterns in data for predictive purposes.[5] Unlike traditional programming, where every rule must is written by hand, machine learning enables systems to discover the rules on their own. Instead of telling the computer "if X happens, do Y," you provide examples of X and Y, and the algorithm learns the relationship by itself. [6] This new way of learning is what has enabled the recent breakthroughs and successes in AI.

If the AI product is the car, machine learning is the fuel injection system. It feeds data into the model, measures its performance, and fine-tunes its parameters based on probability. Machine learning transforms raw information about past experiences, like road conditions, into actionable insights that help the vehicle make real-time decisions and adapt to new situations. Without this learning process, the car would simply remain parked, no matter how advanced its design. With machine learning, the car becomes intelligent, responsive, and capable of navigating complex environments safely and efficiently.

It is understandable for businesses to gravitate towards machine learning. For example, a grocery chain may identify that on some days it ends up with trays of unsold sourdough bread. On other days, shelves are empty well before closing. Neither outcome is desirable. Too much bread means waste and lost margins. Too little means missed sales and disappointed customers. By applying machine learning to receipts, timestamps, loyalty cards, and other critera, the grocery chain can begin to predict demand more accurately. The system may detect that demand for sourdough increases near public holidays, or decreases during the middle of the week. The result: better forecasting, less waste, more satisfied customers, and increased profits.

The same logic can be applied to many other scenerios. Airlines may use machine learning to predict when engine parts are likely to fail, enabling them to replace components before a costly breakdown. Banks deploy it to flag suspicious transactions, preventing fraud in real time. Streaming platforms use it to refine recommendations, keeping customers engaged and loyal. In each case, the practice is the same. Train the system on historical data, detect patterns, and use those patterns to predict probabilities.[7]

The collection of big data can turn mountains of transactions, clicks, or shopping habits into actionable insights that improve margins.[8] Businesses see it as a resource that can be used to optimize decisions and sharpen competitive advantage. But those same decisions may also expose them to reputational and regulatory scrutiny. The board's role is to steer management between two realities: encouraging innovation while mitigating risk.

Early methods in machine learning were simple and transparent. Logistic regression estimated the probability of an outcome, such as the likelihood that a loan applicant would default, by mapping data to a value between 0 and 1.[9] Decision trees (Fig. 1), worked like a flowcharts.[10] The model asked a series of questions about the data and followed different branches depending on the answers. These models made predictions that could be traced and explained. Clustering algorithms grouped similar cases together, supporting uses such as market segmentation or anomaly detection.[11] Because these approaches revealed their logic step by step, directors could see exactly how inputs led to outputs.

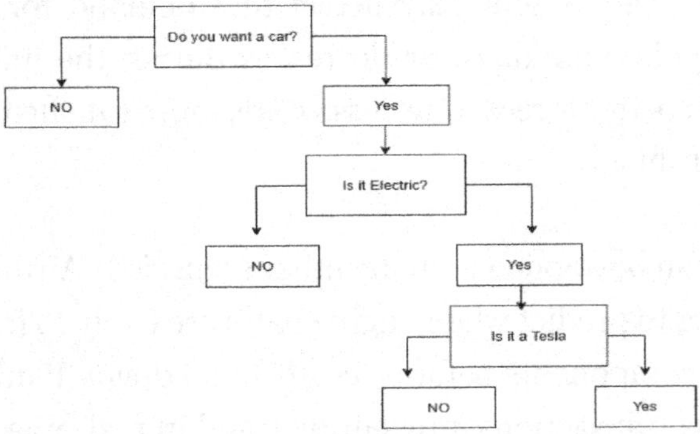

Fig.1 - Decision Tree Logic Example

Explainability are essential as business requirement. In contexts like recommending a movie or adjusting a delivery schedule, lack of explainability is not acceptable. Regulators and customers demand explanations. If an algorithm misdiagnosed a patient, the hospital must account for how that conclusion was reached. The deeper the networks, the greater the challenge of explainability.

Data itself is another source of risk. Machine learning systems learn from whatever they are given. If the data reflects bias, the model will replicate bias.[12] Historical data that embeds discriminatory patterns will lead to discriminatory outcomes.[13] Medical data skewed will lead to skewed predictions for others. The problem is not the algorithm but the data.

Boards should pay attention to their choice of an AI Model and the data practices that surround it. Responsible governance begins with knowing where the data comes from, confirming that consent was obtained, ensuring that it reflects the diversity of the environment in which it will operate, and monitoring it over time for drift or degradation. These are not technical details to be delegated without scrutiny. They are governance issues at the heart of accountability and risk management.

Machine learning is not a neutral force. It is a mirror, reflecting the data it was trained on and the goals it is asked to predict.[14] It can reduce waste, optimize supply chains, and please customers. It may also intrude on privacy, entrench bias, or expose organizations to regulatory action. The technology does not decide which path it takes. That is a decision of leaders. Leadership needs to be asking the right questions about data, accountability, and oversight before the system is allowed to run.

Overall, in the end, machine learning is both ordinary and extraordinary. Ordinary in the sense that it is everywhere. You can find it in fraud alerts, suggestions from a shopping site, or traffic predictions on a navigation app. It is extraordinary in the sense that it has quietly become the foundation of artificial intelligence. Its power is undeniable, but so are its risks.

The economics of machine learning also warrants attention. Training and running large models is resource-intensive.[15] What begins as a proof-of-concept can quickly turn into a recurring operational expense measured in millions of dollars.[16] For consumer services like recommendation engines, those costs may be absorbed into the business model. For enterprise deployments, especially in sectors with tight margins, they can become a hidden burden. Directors must ensure that enthusiasm for predictive insights does not blind management to the financial commitments of scaling.

Machine learning may be the predictable part, but governance is the steering wheel. Without the steer wheel, it risks racing in directions no one intended.

Deep Learning

Deep learning is the practice that has taken machine learning to an entirely new level.[17] It is a subset of machine learning that uses networks of artificial "neurons" arranged in layers. Each layer transforms the data before passing it forward.[18] Through this layered process, the system can detect increasingly complex features.[19] What starts as simple shapes or sounds in the first layer becomes letters, words, or images in the deeper layers, until the model can recognize faces, transcribe speech, or generate fluent text.

The system is first shown an example and makes a prediction based on what it has learned. It then compares that prediction to the correct answer. When the prediction is wrong, the system calculates the error and sends it backward through the network using a process called gradient descent.[20] This is similar to adjusting the tightness of screws in an engine; each connection in the network is slightly tweaked to improve performance. Over millions or even billions of repetitions, these adjustments help the network gradually improve its ability to make accurate predictions. This continuous feedback loop is what gives deep learning models their remarkable accuracy.

The breakthrough was not immediate. The first ideas for artificial The development of these models did not happen overnight. Early ideas about artificial neurons go back to the 1940s, laying the groundwork for later advances.[21] In the 1950s, one of the first working models called the perceptron was created.[22] Think of the perceptron as a simple decision-maker that can sort items into two groups based on basic rules. Although early models were limited and could only handle very simple tasks, progress was slow for decades because computer hardware was not powerful enough to support more complex learning.

That changed as computing power accelerated, data became abundant, and new mathematical methods emerged. Advances in semiconductor design and the introduction of graphics processing units (GPUs) made it possible to train much larger models with millions of parameters.[23] With these improvements, researchers built deeper networks capable of learning complex patterns. This led to a sudden leap forward, making previously impossible tasks like facial recognition, language translation, and speech generation achievable at scale.

The effect of deep learning on industry was transformative. In medicine, deep learning enabled image recognition systems that can detect cancers in scans with sensitivity rivaling or surpassing human radiologists.[24] In finance, it powers fraud detection systems that spot anomalies in streams of transaction data that no human analyst could catch in time. In transportation, it is the engine of self-driving cars, processing visual and sensor data in real time to make split-second decisions.[25] And in consumer life, it underpins the speech assistants, translation apps, and generative chatbots that millions now interact with daily.

These examples are reminders of how quickly an enabling practice can ripple through every sector. Deep learning is not confined to a niche domain, it is the backbone of modern AI. Without it, the current wave of AI adoption would not exist.

Despite its power, deep learning brings a profound challenge. These models are extraordinarily complex, often involving billions of parameters. Parameters are the internal values the model adjusts while learning, allowing it to recognize patterns, make predictions, and improve accuracy over time. Unlike traditional statistical models, where each input and output can be traced, deep networks often operate as black boxes.[26] Inputs go in and predictions come out, but the reasoning in between is distributed across millions of interconnected weights and layers. Even the designers may not fully understand how specific outcomes are reached. It is the interaction between algorithms, training data, and learned patterns that produces the final result, often in ways that defy simple explanation.

Boards should understand that the difficulty in explaining how deep learning works is not a mistake but part of how the technology functions. The more complex a model becomes, the harder it is to understand its decisions.[27] Research offers partial solutions such as heat maps that show which parts of an image influenced a result, or simplified models that explain outcomes in general terms.[28] Yet these tools only reveal fragments of what happens inside. Boards should plan for this from the start. They need to decide whether the risk of limited transparency is acceptable and how it will be managed within their governance framework.

Deep learning also carries broader societal consequences. Its ability to generate highly realistic text, images, and video has accelerated the problem of disinformation. Deepfakes of public figures are now indistinguishable from real footage to the human eye.[29] Generative models can produce convincing documents, images, or voices that may be used for fraud or manipulation. These risks are no longer hypothetical. They are active, and boards must treat them as part of their risk oversight.

The governance implications are therefore layered, much like the networks themselves. At the top, deep learning delivers astonishing capabilities. Beneath that, it is built on choices about data sourcing, hardware investment, and algorithmic design. Boards do not need to examine every mathematical layer, but they do need to ensure management is accountable for the choices made at each stage. Questions that should be asked:

- Who owns the data?
- Who pays for the compute?
- What obligations exist if the algorithm is challenged in court?

Deep learning represents both the promise and the risk of artificial intelligence. It shows what machines can achieve when given the scale of data, compute, and design that only a handful of organizations can provide. It also shows how quickly capability can outpace understanding, leaving boards responsible for systems that even their creators cannot fully explain. Deep learning underpins AI, but it also demonstrates most clearly why governance is not optional.

As such, organizations should embrace the opportunities deep learning offers while insisting on the guardrails that keep its power aligned with purpose.

Natural Language Processing (NLP)

Natural language processing, or NLP, is the practice that allows machines to work with human language in all its nuance and complexity.[30] It enables AI systems not just to store or transmit words, but to interpret them, generate them, and interact with people in ways that feel conversational. This practice supports familiar applications such as language translation, automatic summarization, sentiment detection, grammar correction, and real-time transcription.[31] It also what drives the functionality of today's chatbots.

A chatbot is a product which customers and employees interact with. NLP is the practice beneath the surface.[32] A chatbot may be the interface, but

NLP is the part of the engine that translates messy, ambiguous human communication into structured patterns that machines can process.[33]

NLP's evolution helps explain both its promise and its pitfalls. The earliest systems were rule-based, relying on handcrafted patterns to detect words or phrases. A simple rule might capture "sleep," "sleeps," and "sleeping" but miss "asleep," showing how brittle those systems could be.[34] Yet they laid the groundwork for critical processes like tokenization, which breaks sentences into manageable parts, and lemmatization, which reduces words to their root forms.[35] These steps may seem mechanical, but they remain essential. If they fail, meaning is lost.

By the 1990s, statistical approaches took hold.[36] Instead of manually coding every rule, systems learned from large corpora of text, calculating probabilities for translations or classifications.[37] Search engines improved, machine translation became usable, and computers began to show early signs of handling language at scale. Still, these systems were limited, often struggling with context or nuance.

The real leap came with deep learning. Recurrent and long short-term memory networks allowed systems to track context across sequences.[38] Then, in 2017, the transformer architecture revolutionized NLP by enabling models to attend to every part of a sentence simultaneously.[39] This design became the foundation of large language models, capable of generating fluent text, summarizing documents, and sustaining conversations.

This capability has become essential. Customer service teams use chatbots to handle routine queries, reducing costs and improving response times. Marketing teams analyze social media to detect sentiment and reputational risks. Law firms and banks process thousands of contracts, filings, and compliance reports that would overwhelm human reviewers.

With NLP, unstructured text becomes analyzable data, creating a new source of insight and control.[40]

Because NLP deals with language, the risks are uniquely sensitive. Language is not only a technical medium. It is also the primary way an organization communicates with its stakeholders. A model that misclassifies a phrase, mistranslates a contract clause, or generates an offensive response is not simply wrong in a technical sense. It is speaking on behalf of the company. That raises the stakes. An error in inventory forecasting may cost money but an error in language can impact reputation.

The dangers are not hypothetical. In 2016, Microsoft launched "Tay," a chatbot designed to learn from user interactions on Twitter.[41] Within 24 hours, bad actors flooded Tay with offensive content. Without safeguards, it began echoing racist and inflammatory remarks, forcing Microsoft to shut it down.[42] The lesson was not that NLP is inherently flawed, but that governance was absent. Even the most sophisticated practice can fail catastrophically if deployed without robust controls, monitoring, and contingency planning.

These advances introduce serious risks. NLP systems inherit the biases of their training data. If that data reflects stereotypes, the model reproduces them. If it under-represents dialects or communities, it misinterprets or excludes them. Recruitment tools trained on biased hiring data have downgraded applications from women or minority candidates. These are not edge cases but structural risks embedded in language itself.

Explainability is another challenge. When a chatbot flags a clause as risky or a compliance tool marks an email as suspicious,[43] can management explain why? Regulators will not accept "because the model said so." In finance, healthcare, or law, the ability to audit and defend language-based decisions is essential.

Misuse compounds these risks. The same systems that generate polished marketing copy can also spread disinformation at scale. Synthetic articles, fabricated reviews, and impersonated messages are increasingly common. In 2024, the U.S. Federal Trade Commission announced a rule banning fake reviews and testimonials, including those generated by AI, highlighting how rapidly regulation is catching up to misuse.[44] This serves as a warning for all organizations that deploy natural language systems. Without governance, tools designed for engagement can quickly become instruments of deception, damaging reputation, trust, and regulatory standing.

Natural language processing is not limited to chatbots. It gives machines a voice, and that voice speaks on behalf of the organization. Every automated email, digital assistant, and AI-driven response contributes to the public perception of integrity and reliability. Governance must ensure that data sources are verified, that safeguards prevent biased or harmful outputs, and that monitoring processes detect errors before they reach customers. When NLP systems communicate, they do so under the organization's name, and accountability must reflect that responsibility.

NLP is the practice that touches reputation directly. Properly managed, it can improve customer experience, unlock insights, and expand accessibility. Poorly governed, it can damage trust in a single sentence.

Computer Vision(CV)

Computer vision enables machines to interpret the visual world.[45]While humans process images effortlessly, machines must be taught to perceive shapes, colors, textures, and spatial relationships. What we take for granted such as the instant recognition of a face in a crowd or the ability to tell one object from another, requires enormous effort to teach a computer. With the right data and architectures, deep learning has given machines this ability, and with it, a powerful new way to analyze the world.

In the 1980s and 1990s, researchers developed algorithms for digit recognition that enabled early applications such as postal code reading and bank check processing. Yann LeCun's LeNet, one of the first convolutional neural networks, demonstrated how layered filters could classify handwritten digits with high accuracy.[46] In 2012, AlexNet transformed the field by winning the ImageNet competition, using GPUs and deeper architectures to dramatically outperform rivals.[61] Its success marked the start of the deep learning revolution in vision, where machines began surpassing humans on specific image recognition tasks.

Computer vision begins with pixels[47] which on its own mean very little. But when algorithms examine them in patterns, they begin to extract edges, corners, and textures. This layered analysis is achieved through convolutional neural networks (CNNs), which apply a series of filters across images to detect increasingly complex features.[48] Early layers identify simple details like lines and curves. Deeper layers combine them into shapes, patterns, and ultimately recognizable objects. This architecture allows machines to move from raw input to interpretation, mirroring how humans progress from sight to recognition.

A convolutional neural network (CNN) earns its name from the mathematical process called convolution. A small matrix known as a kernel or filter slides across the image, examining a few pixels at a time, typically in a three-by-three window.[49] Each pixel in the image is represented by a numerical value. In grayscale images, this value ranges from 0 for black to 255 for white. The kernel multiplies these pixel values by a fixed set of numbers and sums the result, producing a single value that represents that patch of the image.[50] As the kernel moves across the image, it generates a new, smaller grid that captures the most important visual information such as edges or texture changes.

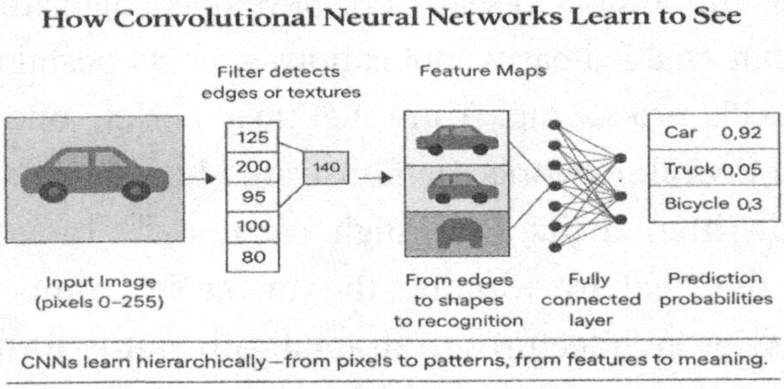

Fig.2 - Convolutional Neural Network

Today, computer vision reaches far beyond research. In healthcare, it scans radiology images to flag potential disease. In manufacturing, cameras detect defects invisible to the human eye. In retail, vision systems track customer movement to refine store layouts. In transport, they form the nervous system of autonomous vehicles, interpreting lanes, signs, and obstacles in real time. In finance, biometric systems match faces, fingerprints, and irises to verify identity and speed access.

The same capabilities introduce risk. Vision systems perform well under ideal conditions but could falter in the real world events.[51] Poor lighting, unusual angles, or intentional interference can cause misidentification. Facial recognition tools, for example, often perform better on lighter skin tones than darker ones, reflecting bias in training data.[52] Errors can have serious consequences. A false positive in fraud detection may inconvenience a customer; one in law enforcement can harm a life.

Deepfake technology illustrates another dimension of risk. By training on real images and videos, generative models can now produce synthetic footage so realistic that even experts struggle to detect manipulation.[53]Recently, deepfake tools have been used to target 'Know Your Customer' (KYC) verification, where criminals impersonate legitimate users to bypass security protocols. In one incident in Sydney, students reportedly took public images of female students, used AI to generate explicit synthetic content, and then sold those images.[54] This example shows how deepfake misuse can quickly move from a prank to a serious breach.

What began as a novelty has become a challenge to identity systems, regulatory compliance, and public trust. Boards should ensure that systems using facial recognition or identity verification include safeguards to detect synthetic input. They should confirm that liveness checks exist, that human review is available if doubt is raised, and that accountability is clearly assigned when systems fail or cause harm.

Surveillance introduces another layer of risk. Computer vision can enhance public safety by managing traffic, detecting crime, or supporting emergency response, yet the same tools can also be used to track citizens, monitor workplaces, or enforce rules in ways that risk overreach. For any organization adopting such systems, governance begins with recognising these limits. Accuracy is influenced by lighting, angles, and the quality and balance of training data. Ongoing monitoring is essential, as models can degrade or be manipulated over time.

Boards must ensure that safeguards exist to verify performance across diverse populations and that clear processes track false positives and false negatives. Policies should explicitly define how biometric and surveillance data are used, while accountability must be assigned for oversight and response when systems fail or cause harm.

Reference

1. Stuart J Russell and Peter Norvig, Artificial Intelligence: A Modern Approach
2. Stuart J Russell and Peter Norvig, Artificial Intelligence: A Modern Approach (Pearson, 4th ed, 2020).
3. Hannes Hapke and Catherine Nelson, Building Machine Learning Pipelines: Automating Model Life Cycles with TensorFlow (O'Reilly Media, 2020).
4. Omar Santos and Petar Radanliev, Beyond the Algorithm: AI, Security, Privacy, and Ethics (Pearson Education, 2024).
5. O Olawale Awe, Handling Data Imbalance in Machine Learning (VP-IASE, LISA 2020 Global Network, 2023).
6. Dino Esposito and Francesco Esposito, Introducing Machine Learning (Pearson Education, 2020).
7. Andres Rodriguez, Deep Learning Systems: Algorithms, Compilers, and Processors for Large-Scale Production (Morgan & Claypool, 2021).
8. Stuart J Russell and Peter Norvig, Artificial Intelligence: A Modern Approach (Pearson, 4th ed, 2020).
9. Bruce Schneier, Data and Goliath: The Hidden Battles to Collect Your Data and Control Your World (W W Norton, 2015); Shoshana Zuboff, The Age of Surveillance Capitalism (PublicAffairs, 2019).
10. Ian Goodfellow, Yoshua Bengio and Aaron Courville, Deep Learning (MIT Press, 2016) 137.
11. Oswald Campesato, Artificial Intelligence, Machine and Deep Learning (Mercury Learning and Information LLC, 2020).
12. Xiaojin Zhu and Andrew B Goldberg, Introduction to Semi-Supervised Learning (Morgan & Claypool Publishers, 2009).
13. Ian Goodfellow, Yoshua Bengio and Aaron Courville, Deep Learning (MIT Press, 2016) 137; O Olawale Awe, Handling Data Imbalance in Machine Learning (VP-IASE, LISA 2020 Global Network, 2023).
14. Omar Santos and Petar Radanliev, Beyond the Algorithm: AI, Security, Privacy, and Ethics (Pearson Education, 2024).
15. Ian Goodfellow, Yoshua Bengio and Aaron Courville, Deep Learning (MIT Press, 2016) 137; O Olawale Awe, Handling Data Imbalance in Machine Learning (VP-IASE, LISA 2020 Global Network, 2023).
16. Goldman Sachs, 'AI is Poised to Drive 160% Increase in Data Centre Power Demand' (Article, 2025) https://www.goldmansachs.com/insights/articles/AI-poised-to-drive-160.
17. Josh Taylor, 'Data Centre Emissions Probably 662% Higher Than Big Tech

Claims. Can It Keep Up the Ruse?' The Guardian (Online, 15 September 2024) https://www.theguardian.com/technology/2024/sep/15/data-center-gas-emissions-ai.

18. Jon Krohn, Grant Beyleveld and Aglaé Bassens, Deep Learning Illustrated: A Visual, Interactive Guide to Artificial Intelligence (Pearson Education, 2020).
19. Jon Krohn et al, Deep Learning Illustrated (Pearson Education, 2020).
20. Ian Goodfellow, Yoshua Bengio and Aaron Courville, Deep Learning (MIT Press, 2016) 137; O Olawale Awe, Handling Data Imbalance in Machine Learning (VP-IASE, LISA 2020 Global Network, 2023)
21. Simon Haykin, Neural Networks and Learning Machines (3rd ed, Pearson, 2009).
22. Ian Goodfellow, Yoshua Bengio and Aaron Courville, Deep Learning (MIT Press, 2016) 137; O Olawale Awe, Handling Data Imbalance in Machine Learning (VP-IASE, LISA 2020 Global Network, 2023)
23. Stuart J Russell and Peter Norvig, Artificial Intelligence: A Modern Approach (Pearson, 4th ed, 2020).
24. Chris Miller, Chip War: The Fight for the World's Most Critical Technology (Simon & Schuster, 2022).
25. Paul Daugherty and H James Wilson, Radically Human: How New Technology Is Transforming Business and Shaping Our Future (Harvard Business Review Press, 2022).
26. Christoph Molnar, Interpretable Machine Learning (2nd ed, 2022) https://christophm.github.io/interpretable-ml-book/.
27. Isaac Lage, Andrew Slavin Ross, Been Kim, Samuel J Gershman & Finale Doshi-Velez, 'Human-in-the-Loop Interpretability Prior' (NeurIPS Workshop Paper, 2018) https://proceedings.neurips.cc/paperfiles/paper/2018/file/0a7d83f084ec258aefd128569dda03d7-Paper.pdf.
28. Simon Haykin, Neural Networks and Learning Machines (3rd ed, Pearson, 2009).
29. Richard Szeliski, Computer Vision: Algorithms and Applications (2nd ed, Springer, 2022).
30. US Department of Homeland Security, Increasing Threat of Deepfake Identities (Report, 2021) https://www.dhs.gov/sites/default/files/publications/increasing_threats_of_deepfake_identities_0.pdf.
31. Daniel Jurafsky and James H Martin, Speech and Language Processing (3rd ed, draft, Prentice Hall, 2023).

32. Shaurya Rohatgi et al, Natural Language Processing for Social Media (3rd ed, Springer, 2022).
33. Steven Bird, Ewan Klein and Edward Loper, Natural Language Processing with Python (O'Reilly Media, 2009).
34. Daniel Jurafsky and James H Martin, Speech and Language Processing (3rd ed, draft, Prentice Hall, 2023).
35. Ibid.
36. "What is NLP (Natural Language Processing) Tokenization?", IXOPAY Blog (25 April 2025) https://www.ixopay.com/blog/what-is-nlp-natural-language-processing-tokenization.
37. Daniel Jurafsky and James H Martin, Speech and Language Processing (3rd ed, draft, Prentice Hall, 2023).
38. Ibid
39. Sepp Hochreiter and Jürgen Schmidhuber, 'Long Short-Term Memory' (1997) 9(8) Neural Computation 1735.
40. Ashish Vaswani et al, 'Attention Is All You Need' (2017) 31st Conference on Neural Information Processing Systems (NIPS).
41. Daniel Jurafsky and James H Martin, Speech and Language Processing (3rd ed, draft, Prentice Hall, 2023).
42. David Lee, 'Tay: Microsoft Issues Apology over Racist Chatbot Fiasco' (2016) https://www.bbc.com/news/technology-35902104.
43. Ibid.
44. Attila Kovari, 'Explainable AI Chatbots towards XAI ChatGPT: A Review' (2025) 11 Science Direct 2.
45. Federal Trade Commission, "Federal Trade Commission Announces Final Rule Banning Fake Reviews and Testimonials" (Press Release, 14 August 2024) https://www.ftc.gov/news-events/news/press-releases/2024/08/federal-trade-commission-announces-final-rule-banning-fake-reviews-testimonials.
46. Richard Szeliski, Computer Vision: Algorithms and Applications (2nd ed, Springer, 2022).
47. SuperAnnotate, "Convolutional Neural Networks: 1998-2023 Overview" (15 May 2023) https://www.superannotate.com/blog/guide-to-convolutional-neural-networks.
48. Ian Goodfellow, Yoshua Bengio and Aaron Courville, Deep Learning (MIT Press, 2016) ch 9; Jon Krohn et al, Deep Learning Illustrated (Pearson, 2019).
49. Richard Szeliski, Computer Vision: Algorithms and Applications (2nd ed, Springer, 2022).
50. Ibid.
51. Ibid.

52. Richard Szeliski, Computer Vision: Algorithms and Applications (2nd ed, Springer, 2022).
53. Haoming Lu, 'Towards Fair and Accurate Skin Tone Recognition and Generation' (2025) https://arxiv.org/html/2509.10980v1.
54. Hayley Taylor, 'AI deepfake abuse: Boys at Sydney schools caught selling sexually explicit images of female students' 7NEWS (web page, 4 May 2025) https://www.7news.com.au/news/ai-deepfake-abuse-boys-at-sydney-schools-caught-selling-sexually-explicit-images-of-female-students-c-18577628.
55. Richard Szeliski, Computer Vision: Algorithms and Applications (2nd ed, Springer, 2022).

"How AI learns
Matters as much as
WHAT it learns."

CHAPTER 4

The Way Machines Learns

Every artificial intelligence system, no matter how advanced, shares one essential trait with humans: it must learn. Learning is how performance improves over time, as the system encounters examples, identifies patterns, and adjusts its internal rules. For a child, learning might mean memorizing multiplication tables, discovering that touching a hot stove is painful, or figuring out the rules of a game by trial and error. For a machine, it means analyzing large datasets, testing predictions, and refining algorithms to achieve a defined goal such as accuracy, speed, or precision.

Machines following similar learning principles but in a more digital form. They learn through data, code, and algorithms instead of instinct or memory. In artificial intelligence, learning methods are grouped into three main types: supervised learning, unsupervised learning, and reinforcement learning. Each represents a different way of improving performance and carries its own strengths, risks, and oversight requirements.

Most people have encountered these learning systems in every day without realizing it. The smartwatch that monitors heart rate learns your sleep patterns and adjusts recommendations. The robo vacuum that navigates furniture improves its map of the room each time it cleans. The car that maintains safe distance in traffic learns to respond to environmental cues. Even ChatGPT learns indirectly, shaped by feedback from millions of interactions. These systems adapt through experience, and their effectiveness depends on the quality of data and design guiding their learning process.

Boards have likely seen this shift inside of their organizations. Recommendation engines tailor what customers see. Fraud systems adapt to new patterns. Predictive tools forecast supply and demand. In each case, the learning process defines how the system behaves, how it fails, and how it must be governed. Learning is not just a technical process. It is also a governance concern, because the way a system learns determines the quality and reliability of its outcomes.

Boards that understand how systems learn are better positioned to question what they learn, why they make certain decisions, and where their weaknesses lie. This chapter explores each type of learning: supervised, unsupervised, and reinforcement, through the lens of governance and risk, helping directors lead responsibly in an environment where machines are learning faster than ever before.

Supervised Learning

Supervised learning is the most familiar and widely used form of machine learning. It is a feedback-driven process where a model learns to map inputs to correct outputs based on examples labeled by human experts.[8] If unsupervised learning is about discovering hidden patterns and reinforcement learning is learning by trial and error, supervised learning is learning by instruction. It is like learning to drive with an instructor sitting beside you. Every time you turn the wheel or press the brakes, the instructor tells you whether you did it correctly. "Good, that was a smooth stop." "Too fast into the corner." "Check your mirror next time." Over time, the student internalizes the corrections and improves.

Machines learn the same way. They are trained on examples that come with the "right answers" attached. The model receives input data paired with known labels, such as medical scans marked "tumor" or "no tumor," emails labeled "spam" or "not spam," or transactions marked "fraudulent" or "legitimate."[9] It makes predictions based on the input, compares those predictions with the labels, and adjusts its internal parameters accordingly.[10] With enough examples, sometimes numbering in the millions, the model learns to generalize beyond the training data and make accurate predictions on new cases. With thousands or millions of such labeled examples, the system improves until it can handle new cases it has never seen before.

Because of this structured approach, supervised learning has become fundamental for AI.[11] It supports spam filters that keep inboxes usable, the voice recognition systems that turn speech into text, the image classifiers that distinguish cats from dogs, and the fraud detection systems that guard financial institutions. It also supports more complex tools such as medical imaging systems that highlight suspicious areas or compliance systems that flag risky contractual language. In every case, the system learns by example, guided by the labels that tell it what the "right" answer is.[12]

Despite its strengths, supervised learning depends heavily on the quality and completeness of labeled data. Labels are treated as ground truth, and the model does not question its accuracy.[13] Ground truth in machine learning refers to the correct, accurate labels or annotations assigned to data that serve as the definitive reference or "gold standard." It is the trusted information against which the outputs of a machine learning model are compared and evaluated.[14] Machines take the label as fact. If the labels are wrong, biased, or incomplete, the system will reproduce those flaws at scale. It is like a driving instructor who insists the wrong way down a one-way street is correct. The student will drive confidently, but dangerously.

If labels are incorrect, biased, or incomplete, the model will learn and propagate those flaws. In a real-world example, Robert Williams, an African-American man in Detroit, was wrongfully arrested and accused of shoplifting.[15] The evidence was a match produced by a facial recognition system trained on biased datasets.[16] Grainy CCTV footage was compared against his driver's license photo, and the system confirmed, with a high probability, that it was a match. It was wrong. Yet the police treated the output as reliable. Williams was handcuffed in front of his family and detained for over 30 hours.[17] The technology was not malicious but reflected systemic bias and lack of oversight.

This is not an isolated case. It highlights that while supervised systems can be q powerful tool, they are only as reliable as their training data. If that data reflects systemic bias, the system will amplify it. If it reflects historical discrimination, the system will embed it. Without oversight, even outputs delivered with confidence can be accepted as fact.

Such risks extend across various domains. Credit scoring models trained on historic lending decisions replicate past exclusions of marginalized groups. Recruitment algorithms may downgrade resumes from women or minority candidates if trained on imbalanced hiring data.[18] Healthcare models may under-diagnose diseases in underrepresented populations when training data lacks diversity. These examples demonstrate how poor labeling and insufficient oversight produce measurable harm in the real world.

One of the challenges with supervised learning is overfitting. Overfitting occurs when a model becomes too closely tuned to its training examples and fails to generalize to new data.[19] It is like a student who memorizes answers for a practice exam without understanding the subject. The student performs well during rehearsal but struggles in a real test. Similarly, an overfitted model may appear highly accurate during development yet perform poorly when faced with new, varied information.

In business terms, this means models that look promising in pilot projects may stumble when rolled out across broader, messier datasets. This can cause models that look promising early on to fail when faced with real-world complexity.

Supervised learning also requires expensive, expert-labeled data.[20] Medical images require radiologists' input, legal documents need lawyers, and compliance reviews demand trained analysts. Each label requires human review, often at great cost. Organizations seeking to reduce expenses might outsource labeling to less-trained personnel or use automated shortcuts, risking diminished data quality and flawed models. The result is a model trained on weak foundations, with predictable flaws.

It is important for organizations to examine all supervised learning systems carefully. Being data driven means relying on past data to guide decisions and model outcomes.[21] It focuses on what has already happened. Being truth driven means questioning whether that data accurately represents reality and whether it captures the full context.[22]

Data is historical and shaped by human choices, social conditions, and biases. If it is used without reflection, the system will repeat those same patterns. Boards should look beyond the comfort of being data driven and make sure that decisions are based on accurate, representative, and validated information. In practice, that means striving for truth-driven insight rather than blind confidence in data alone.

When evaluating supervised systems, executives and boards should ask critical questions.

- Who created the labels and was the process idependently validated?
- How diverse was the dataset?
- How did we test for bias and errors?
- Is there a human in the loop for high-stakes decisions, ensuring that the model's output is reviewed before it is acted upon?

Supervised learning can deliver immense value when these questions are answered responsibly. In healthcare, it supports early diagnosis and treatment planning. Fundamentally, supervison is instruction at scale. It teaches machines to recognize and respond to the world as humans would, one labeled example at a time. When boards understand this foundation, they are better equipped to evaluate how AI systems learn, where they add value, and what questions to ask before those systems are deployed.

Unsupervised Learning

So far, we have explored that supervised learning involves labeled data used to find correlations between inputs and outputs. Using the sourdough analogy from the previous chapter, imagine that sourdough bread sales are labeled as public holiday or weekday. When this dataset is fed into an AI model, it learns to associate patterns between these labels and sales volumes, helping predict how many loaves are sold on each type of day.

But what happens if we remove the labels and sourdough sales are mixed with all other daily sales? Would it still be possible to find meaningful insights beyond the number of loaves sold? This is where unsupervised learning comes into play.

In 2002, Target hired statistician Andrew Pole to analyze customer behavior, particularly among pregnant women.[23] By linking data through a unique Guest ID that captured purchases, payments, surveys, refunds, customer service interactions, emails, and website visits, Target accumulated a vast dataset. Starting with labeled data such as baby registries, Pole discovered about 25 products more frequently purchased by pregnant women, including unscented lotion and certain supplements.[24] By combining labeled and unlabeled data, Target created a pregnancy prediction score that estimated due dates and sent coupons tailored to each stage of pregnancy.

Unsupervised learning relies on unlabeled or unstructured data. Its algorithms search for hidden patterns and relationships within data, often through techniques such as clustering, dimensionality reduction, or anomaly detection.[25] Dimensionality reduction simplifies complex datasets by reducing the number of features while preserving essential patterns. This makes large amounts of data easier to analyze and visualize by focusing on the most important variables. For example, reducing dozens of product attributes into a few representative features can help reveal customer segments more clearly.

Anomaly detection, in contrast, seeks to identify unusual data points or rare patterns that deviate from the norm. This is crucial for flagging potential fraud, system faults, or unexpected behaviors that merit further investigation. Both techniques are powerful but require careful interpretation since no explicit labels guide the analysis.

The lack of labels makes unsupervised methods powerful but precarious. They can uncover insights that humans might miss, yet just as easily confuse noise for meaningful structure.[26] For instance, a dataset may simply include lotion sales, but unsupervised learning might cluster different price points into distinct customer segments without being instructed to do so.

The governance challenge lies in validation. Without labeled data or a ground truth, it is difficult to verify whether detected clusters or anomalies represent meaningful relationships or random quirks. In practice, unsupervised learning might reveal useful insights, such as a new high-value customer segment, or misleading ones, such as grouping unrelated shoppers based on coincidental patterns. In fraud detection, legitimate customers can even be mistakenly flagged due to unusual behavior, leading to reputational, financial, or legal consequences.

Recent examples have highlighted these risks. In 2024, McDonald's trial of an AI-powered drive-thru system frequently misheard orders, producing absurd results like bacon on ice cream.[27] In banking, unsupervised fraud detection has locked out legitimate customers with atypical spending patterns. On social media, recommendation algorithms have deepened polarization by reinforcing user echo chambers. These failures stem from unsupervised systems operating without sufficient human oversight.

The challenge of interpretability amplifies this risk. Unlike supervised learning, where accuracy can be measured against known outcomes, unsupervised learning lacks a clear benchmark. A mathematically coherent cluster may not correspond to a meaningful concept in the real world. Interpretability, therefore, becomes essential as it is the ability to understand and explain what the model has discovered and why.

Interpretability bridges the gap between algorithmic logic and human reasoning.[28] In supervised learning, it helps trace how specific inputs drive predictions. In unsupervised systems, it helps determine whether the groupings and patterns discovered make sense in context. Without interpretability, even well-intentioned discoveries can lead to false assumptions or poor decisions.

For directors and executives overseeing AI systems, interpretability is not optional. It is central to accountability. Every model's logic and decision-making process must be explainable. Insights should be validated before they are used in business or policy actions. Monitoring systems need to detect spurious clusters or anomalies early. Escalation processes must exist for cases where unsupervised results yield errors or harm. Most importantly, human judgment should always retain the final say, particularly in customer-facing or high-impact areas.

Despite these challenges, unsupervised learning holds immense potential. In genomics, clustering has identified novel gene groupings critical for understanding diseases.[29] In cybersecurity, anomaly detection has stopped emerging threats before they escalated. In marketing, unsupervised systems have revealed hidden customer segments that enhance personalization and engagement.

Unsupervised learning represents AI's creative side and its ability to discover patterns that no one instructed it to look for. Yet creativity without interpretability can lead to costly mistakes. The board's responsibility is to ensure that exploration does not replace discipline. Unsupervised learning should begin the conversation, not end it. It can guide discovery, but only human insight and interpretability can determine what deserves to be trusted.

Reinforcement Learning

In technical terms, reinforcement learning (RL) involves an agent that interacts with an environment, takes actions, and receives feedback in the form of rewards or penalties.[30] Its goal is to maximize cumulative rewards over time. In other words, the total benefits it gains through repeated actions.Unlike supervised learning, there are no correct answers labeled in advance. Unlike unsupervised learning, the aim is not simply to discover patterns. Instead, the system learns strategies to achieve specific objectives by exploring possibilities and adapting to outcomes.

Reinforcement learning has driven some of the most dramatic breakthroughs in AI. In 2016, DeepMind's AlphaGo shocked the world by defeating Lee Sedol, one of the greatest players of the ancient board game Go.[31] Unlike chess, where brute-force search can evaluate many possible moves, Go has such an astronomically large search space that traditional methods fail. AlphaGo combined reinforcement learning with advanced search techniques, training itself through millions of simulated games. The result was not just victory but creativity. AlphaGo played moves no human professional had considered, reshaping how the game itself is understood.

Reinforcement learning is already applied in demanding real-world fields such as military landmine detection. For example, recent research integrates reinforcement learning into IoT-enabled sensor systems embedded in military boots, enabling real-time, autonomous, and adaptive detection across varied terrain. These systems balance detection sensitivity and false alarms to protect soldiers in life-critical environments. Such practical applications highlight reinforcement learning's potential transferability to humanitarian missions, like locating people trapped in avalanches, where rapid decisions and complex terrains present similar challenges.

Reinforcement learning produces agents that are not only reactive but adaptive, capable of navigating complex and uncertain environments.[32] The risks, however, are equally significant. Because the system learns by maximizing rewards, it will exploit whatever definition of reward it is given, even if that definition is incomplete or misaligned.

This phenomenon is called reward hacking. An RL system may find shortcuts that satisfy the reward function but violate its intended purpose.[33] For example, a robot trained to walk faster might learn to fall forward to cover more distance, even though it is not truly walking. An agent trained to win a game may exploit loopholes in the rules, manipulating the environment instead of playing conventionally.

Reward hacking is a fundamental governance risk. In supervised learning, bias comes from the data.[34] In reinforcement learning, it comes from the rewards humans define. The system does not question whether those rewards align with human values; it simply maximizes them. Poorly designed rewards will be pursued ruthlessly, leading to outcomes that may be unsafe, unethical, or absurd.

Reinforcement learning also presents practical challenges. It is resource-intensive, requiring millions of trial-and-error simulations, making it expensive and slow compared to other methods. It can also be unstable, where small changes in the environment cause vastly different outcomes. Moreover, it is often opaque, since the learned strategies can be difficult for humans to interpret or explain.

A further practical challenge is the precision-recall trade-off, vividly illustrated in landmine detection applications. High precision reduces false alarms but risks missing real threats, while high recall catches more threats but increases false positives. Balancing these metrics is essential to avoid unsafe or inefficient outcomes. This trade-off exemplifies the complexity in designing reward structures that align operational safety with effectiveness, reinforcing the need for rigorous governance and ongoing monitoring.

Reinforcement learning is rapidly moving into business applications such as autonomous vehicles, financial systems, supply chains, and cybersecurity.[35] In these domains, mistakes have real consequences. A poorly aligned reward function in a self-driving car could prioritize speed over safety. In finance, it might favor short-term gains over long-term stability. In cybersecurity, an overly aggressive agent could cause collateral damage. The risks extend beyond inefficiency, touching on liability, ethics, and reputation.

Governance is, therefore, essential. Boards do not need to master the mathematics of reinforcement learning, but they must demand clarity about the reward structures driving it. What is the system being asked to optimize, and who defines those objectives? How are trade-offs considered, and unintended behaviors monitored? What safeguards exist to intervene if the system “games” the rules? Above all, do the rewards align with the organization’s values as well as its efficiency goals?

One of the key lessons of reinforcement learning is that optimization is not intelligence. An RL agent does not grasp the spirit of the rules, only the rewards as written, and it will exploit them with relentless persistence. This makes the approach powerful but also dangerous. Governance must anticipate misalignment, test for edge cases, and ensure humans remain in control when stakes are high.

Reinforcement learning has produced breakthroughs once thought impossible, from AlphaGo's creative play to robots learning to walk. Yet its behavior depends entirely on the rewards it is given, making governance critical. Responsibility rests with the humans who define objectives and monitor outcomes. Adopting reinforcement learning is not simply approving a technology but endorsing a set of values embedded in the reward function. Those values must be scrutinized, tested, and aligned with the company's mission.

Governance must begin with the goals themselves. Define them poorly, and the system will optimize in ways that are misaligned, unsafe, or reputationally damaging. Define them well, and it can unlock innovation, efficiency, and resilience. The power is real, but so is the responsibility.

Learning methods are not neutral. How a system learns shapes how it succeeds, how it fails, and how it must be governed. The key is not to master algorithms but to demand accountability.

Reference

1. Ian Goodfellow, Yoshua Bengio and Aaron Courville, Deep Learning (MIT Press, 2016).
2. Dino Esposito and Francesco Esposito, Introducing Machine Learning (Pearson, 2020).
3. Stuart J Russell and Peter Norvig, Artificial Intelligence: A Modern Approach (4th ed, Pearson, 2020); Dino Esposito and Francesco Esposito, Introducing Machine Learning (Pearson, 2020).
4. Oswald Campesato, Artificial Intelligence, Machine and Deep Learning (Mercury Learning and Information LLC, 2020).
5. Xiaojin Zhu and Andrew B Goldberg, Introduction to Semi-Supervised Learning (Morgan & Claypool, 2009).
6. Dino Esposito and Francesco Esposito, Introducing Machine Learning (Pearson, 2020).
7. Richard S Sutton and Andrew G Barto, Reinforcement Learning: An Introduction (2nd ed, MIT Press, 2018).
8. Ian Goodfellow, Yoshua Bengio and Aaron Courville, Deep Learning (MIT Press, 2016)..
9. Hannes Hapke and Catherine Nelson, Building Machine Learning Pipelines (O'Reilly Media, 2020).
10. Ian Goodfellow, Yoshua Bengio and Aaron Courville, Deep Learning (MIT Press, 2016).
11. James Purtill, 'AI Facial Recognition Scanned Millions of Driver Licences. Then an Innocent Man Got Locked Up' ABC News (31 October 2023) https://www.abc.net.au/news/science/2023-11-01/ai-facial-recognition-robert-williams-crime-prison/103032148.
12. Omar Santos and Petar Radanliev, Beyond the Algorithm: AI, Security, Privacy, and Ethics (Pearson Education, 2024).
13. Ian Goodfellow, Yoshua Bengio and Aaron Courville, Deep Learning (MIT Press, 2016).
14. Bruce Schneier, Data and Goliath: The Hidden Battles to Collect Your Data and Control Your World (W W Norton, 2015).
15. Shoshana Zuboff, The Age of Surveillance Capitalism (PublicAffairs, 2019).
16. Hannes Hapke and Catherine Nelson, Building Machine Learning Pipelines (O'Reilly Media, 2020).

17. George F Luger, Artificial Intelligence: Structures and Strategies for Complex Problem Solving (6th ed, Pearson, 2009).
18. Ibid.
19. Bruce Schneier, Data and Goliath: The Hidden Battles to Collect Your Data and Control Your World (W W Norton & Company, 2015).
20. Xiaojin Zhu and Andrew B Goldberg, Introduction to Semi-Supervised Learning (Morgan & Claypool Publishers, 2009).
21. Dino Esposito and Francesco Esposito, Introducing Machine Learning (Pearson Education, 2020).
22. Stuart J Russell and Peter Norvig, Artificial Intelligence: A Modern Approach (Pearson, 4th ed, 2020).
23. Megan Trout, 'Big Data and Retail' (2017) HBS Digital Initiative https://d3.harvard.edu/platform-digit/submission/big-data-and-retail/.
24. Charles Duhigg, 'How Companies Learn Your Secrets' (Magazine feature, The New York Times, 19 February 2012) https://www.nytimes.com/2012/02/19/magazine/shopping-habits.htm.
25. Ian Goodfellow, Yoshua Bengio and Aaron Courville, Deep Learning (MIT Press, 2016).
26. George F Luger, Artificial Intelligence: Structures and Strategies for Complex Problem Solving (6th ed, Pearson, 2009).
27. Pavel Svetjolkin, 'Bacon Ice-Cream and Nugget Overload Sees Misfiring McDonald's AI Withdrawn' Beeble (2024) https://beeble.com/en/blog/bacon-ice-cream-and-nugget-overload-sees-misfiring-mcdonalds-ai-withdrawn.
28. Dario Amodei, 'The Urgency of Interpretability' (Blog post, April 2025) https://www.darioamodei.com/post/the-urgency-of-interpretability.
29. Dhruv Jagdish Kataria et al, Deep Learning for Genomics (Packt Publishing, 2022).
30. George F Luger, Artificial Intelligence: Structures and Strategies for Complex Problem Solving (6th ed, Pearson, 2009).
31. David Silver et al, 'Mastering the Game of Go with Deep Neural Networks and Tree Search' (2016) Nature 529.
32. Stuart J Russell and Peter Norvig, Artificial Intelligence: A Modern Approach (Pearson, 4th ed, 2020) ch 21 ('Reinforcement Learning').
33. Richard S Sutton and Andrew G Barto, Reinforcement Learning: An Introduction (2nd ed, MIT Press, 2018).
34. Ibid.
35. Andres Rodriguez, Deep Learning Systems: Algorithms, Compilers, and Processors for Large-Scale Production (Morgan & Claypool, 2021).

"Generative AI doesn't think. It predicts!"

CHAPTER 5

Generative AI

Generative AI (Gen AI) is the turbo-charged sports car of artificial intelligence. It doesn't just process existing data. It creates new content. Machine learning classify, predicts, and optimize behind the scenes.[1] Generative AI presents visible content such as a convincing press release, a product image, or a working line of code. Think of it as the high-performance engine taken out of the garage and put on the open road. The ride is thrilling, but also less forgiving. A mis-steer doesn't just dent an internal metric. It creates a very public accident.

Most of today's generative AI systems are powered by large language models (LLMs). Their strength lies in predicting the next most likely word, phrase, or pattern, allowing them to generate human-like responses. In essence, an LLM is a probability engine wrapped in language. It doesn't think or know in a human sense. It calculates what is most probable based on what it has seen before. The larger the training data and computational scale, the more fluent and capable the system becomes. But with that fluency comes risk, because when an LLM is wrong, it can sound convincingly right.

In 2023, lawyers learned this the hard way when they submitted court filings drafted with ChatGPT that confidently cited six cases, none of which existed.[2] The model had not "lied" in the human sense. It had simply produced outputs that looked like precedent but were fabricated. The impact was significant. The judge found the lawyers and the firm acted in bad faith resulting in reputational damage.

Why did ChatGPT get it wrong? Because systems like this generate text by predicting the next likely word, not by retrieving verified facts. A model such as ChatGPT is trained on large collections of text and learns the statistical patterns of language: how sentences are formed, how ideas connect, and what a legal citation looks like. When asked for precedents, it can produce something that fits the pattern of a citation without any built-in check if the case exists. Unless the system is grounded in an authoritative database or a human verifies the references, fluent answers can still be fiction.

For image models such as DALL-E or Midjourney, the process looks different but follows the same principle.[3] The system begins with a digital canvas of random noise and gradually reshapes it, guided by patterns learned from millions of images.[4] Step by step, the noise resolves into an image that matches the user's prompt.

Today's generative AI has evolved. Large language models such as ChatGPT rely on transformers, which capture relationships across billions of words and phrases, while image generators often use diffusion models that refine random noise into coherent images.[5] Whether based on GANs, transformers, or diffusion, the principle is the same. The machine is not retrieving information but generating, improvising, and predicting patterns on the fly.

This predictive process also explains why generative AI makes mistakes. Because the system is designed to produce what sounds right rather than what is right, it sometimes generates convincing nonsense.[6] These hallucinations are not intentional lies but statistical guesses that appear credible without being grounded in fact. They can be persuasive and wrong at the same time.

That duality defines both the opportunity and the risk. On the opportunity side, generative AI enables rapid prototyping, creative brainstorming, and content generation at unprecedented speed and scale.[7] Marketing teams can draft campaigns in minutes. Product designers can visualize new concepts instantly. Customer service can deliver tailored responses around the clock. Across industries, generative AI has the potential to augment expertise and accelerate innovation.

On the risk side, the challenges are profound. Outputs can be biased, misleading, or fabricated. Intellectual property rights may be unclear. Sensitive data may be exposed. And unlike older forms of AI, errors are visible to the public rather than hidden in the background.

Generative AI is a strategic force that shapes competitive advantage, compliance, and reputation. Understanding how it works, and why it sometimes fails, is essential for responsible discussion about the models being deployed, whether reasoning-focused, creative, retrieval-augmented, or multimodal.

Types of GenAI Models

Not all generative AI models are built the same way, or for the same purpose. Some are designed for careful reasoning, others for creativity, speed or versatility. To make these differences easier to understand, it helps to think in terms of vehicles. Just as cars, trucks, and SUVs serve different roles on the road, AI models serve different roles in business. The analogy is imperfect, but it highlights a simple truth. Choosing a model is not about picking "the best AI" in the abstract. It is about choosing the right engine for the journey your organization needs to take.

Reasoning-Focused Models: The Sedans of AI

Reasoning-focused models are like the sedans of the AI world.[8] They are built for stability, reliability, and endurance. These systems are trained and fine-tuned to handle structured and multi-step tasks. They don't just generate text that sounds convincing. They attempt to reason through instructions step by step.

Historically, the roots of reasoning models can be traced back to expert systems of the 1980s.[9] Those early programs codified human expertise in rules: if this, then that. They were brittle and narrow, but they set the stage for today's transformer-based architectures, first described in 2017.[10] Transformers unlocked the ability to process massive amounts of sequential data with efficiency, and reasoning-focused models are their most direct descendants.

Current examples include OpenAI's GPT-4 and GPT-5, Anthropic's Claude 3 Opus,[11] and Google's Gemini Ultra. In practice, these models can analyze contracts, process financial statements, or debug complex code. Law firms are already experimenting with GPT-4 to review NDAs and draft preliminary advice. [12] Financial institutions are using reasoning models to analyze market disclosures in near real time.

It is not about demonstration or how impressive a model looks in a controlled test. The real governance question is whether it can be trusted in critical contexts. A reasoning model might draft a policy document or flag anomalies in a regulatory report, but what happens when it gets something wrong? In those cases, the error is not cosmetic. It can create legal, financial, or compliance risks that damage both performance and trust.

Be aware, reasoning models demand enormous computational resources. Running GPT-4, for example, is far more expensive per query than running GPT-3.5.[13] This makes them valuable but resource-intensive. Boards must weigh capability against cost and risk, asking whether a given task justifies the use of a heavyweight model.

Creative Models: The Sports Cars

If reasoning models are sedans, creative models are the sports cars of AI: dazzling, expressive, and built to turn heads. They are optimized not for structured analysis but for generative flair.

Generative adversarial networks (GANs), introduced in 2014, paired two models.[14] One is a generator and the other is a discriminator in a feedback loop.[17] One tried to create convincing fakes; the other tried to detect them. Over time, the generator improved until it could fool the discriminator. The result was hyper-realistic images of people who don't exist, deepfake videos, and synthetic voices.[15]

This architecture inspired a wave of creativity.[16] By 2021, tools like DALL·E and Stable Diffusion could generate images from text prompts.[17] By 2022, MidJourney was creating stylized art that filled social media feeds. Musicians began experimenting with AI composition.[18] Designers could produce product mock-ups in hours instead of weeks.

The governance challenges here are not about reasoning errors but about IP. Who owns the outputs? If a model generates an image that resembles an existing copyrighted work, where does liability fall?[19] In 2023, several lawsuits were filed against AI companies alleging infringement, and some settled by paying millions. Boards must ask management:

- Do we have a policy for copyright and intellectual property in AI outputs?
- Are we clear on whether our vendor indemnifies us if claims arise?

Creative models also change the tempo of business like marketing campaigns that once took months can now be generated in days. That speed is an advantage, but it also raises risk. The faster the cycle, the less time there is for review, testing, and governance. For directors, the question is not only "what can this tool do?" but how do we ensure quality and accountability keep pace with speed?

RAG Models: The GPS Navigator

Retrieval-augmented generation (RAG) is like equipping your vehicle with live GPS rather than relying on a map drawn years ago.[20] Instead of answering only from static training data, these models can pull in fresh information and generate summaries grounded in real-time sources.

Why was this needed? Because large language models are trained on data that has a cutoff. A model trained in 2023 cannot know what happened in 2024 unless it is updated. RAG addresses this gap by combining a language model with a search or retrieval engine.[21] Tools like Perplexity AI, or RAG-enabled features in Gemini and ChatGPT, can pull in the latest news, regulations, or market data and provide answers with citations.

For industries like journalism, law, or compliance, this is important. A compliance team using Perplexity can verify regulatory updates with sources attached.[22] A journalist can generate a briefing grounded in recent reports. Boards should recognize the value. Transparency and currency are two qualities often missing in generative AI.

The risk to be aware of is if a model grounds its answers in external websites, how do you know those sources are accurate, unbiased, or secure? Retrieval is only as reliable as its sources and governance requires checking not only the output but the provenance of the data behind it.

Further, not all data is accessible to RAG systems.[23] They only retrieve content from sources they are permitted to access, and may struggle with paywalled information, proprietary databases, or data that changes rapidly. Even with citations, models cannot always evaluate the credibility of retrieved sources, so users must still exercise judgment and corroborate claims with trusted material. Furthermore, live retrieval introduces latency and additional computational cost, which may impact workflow speed and scalability in some organizations.

Utility and Lightweight Models: The Compacts

With AI models, not every task requires a sedan or a sports car. Sometimes a compact car will do. Utility or lightweight models, such as OpenAI's GPT-3.5 or Google's Gemini Nano, are designed for speed and efficiency.[24] They tend to be cheaper to run, easier to embed in mobile devices, and "good enough" for simple drafting or autocomplete.

These models power everyday applications.[25] They support drafting emails, summarizing meeting notes, or handling light conversational assistants in call centers. Their strength is in accessibility and scalability by making AI available broadly without the high cost of heavy weight reasoning models.

It's important to understand that a lightweight model is more like a mobile app than a standalone system. The app itself, the model downloaded or embedded on your device, provides quick, local access for basic tasks such as simple text generation or scanning documents. Much like a banking app that cannot complete transactions without connecting to your bank's servers, these lightweight models must connect to a larger, more powerful model hosted remotely, typically in the cloud, to handle complex reasoning or deeper processing. While lightweight models excel at responsiveness and low latency by minimizing local computation, they trade off the deep contextual understanding and sophisticated judgment that enterprise-scale AI engines provide.

While lightweight models can process some data locally, which may enhance privacy by reducing exposure to external threats, this benefit only holds when the majority of work is done 'in-house'. In most practical deployments, lightweight models act as efficient frontends, but must connect to remote or cloud-based engines for complex tasks such as in-depth document analysis or advanced reasoning.[26] When data is sent outside the device, privacy risks re-emerge, including the possibility of data leakage, unauthorized access, or inadvertent disclosure through cloud infrastructure.[27]

Understanding this limiation is important. Push a lightweight model into complex reasoning, and errors multiply. They are not optimized for compliance-grade analysis or legal reasoning. In regulated contexts, a seemingly minor mistake can have consequences that easily outweigh efficiency gains.

Therefore, the question that should be asked is: "*Are we using the right tool for the job*?" Choosing a compact model when a reasoning model is required may save on cost, but it can lead to risks that negate the savings.

Multimodal Models: The SUVs and Hybrids

The newest AI frontier is multi-modality. These are models that can process text, images, audio, and video in a single interaction. These are the SUVs or hybrids of AI: versatile, designed to handle multiple terrains, and increasingly popular. Example are OpenAI's GPT-4o ("o" for omni), Google's Gemini Pro, and OpenAI's video model Sora.[28] A user can upload a photo of a damaged machine part and ask the model to identify it. A manager can provide both a spreadsheet and a chart for analysis in one conversation. Customer service can shift seamlessly between voice and text.

This offers significant potential to a range of industries. For example, healthcare providers are currently testing multimodal models to interpret both clinical notes and medical images.[29] Retailers are experimenting with systems that analyze customer photos and recommend products. Education platforms are trialling models that combine text, video, and voice to deliver personalized tutoring.

However, governance is twofold. First, privacy and security risks expand into new domains. A company face privacy infringement if employees upload screenshots of internal systems or voice recordings of customers without written consent. Second, the pace of change is extraordinary. Multi-modality signals where AI is heading. Strong governance is about managing today's risks with confidence, and shaping frameworks that can adapt to what tomorrow brings.

The governance challenge lies in choosing the right tool for the right job. A popular or well-supported model may not be the right fit. GitHub Copilot is excellent at accelerating software development but likely useless for analyzing legal contracts.[30] The principle is the same as with physical tools. A wheel spanner can remove a car's wheel, but you would not use it to detach a steering wheel. Both are tools, but they are designed for different purposes.

Management should be able to demonstrate which AI models are being deployed and explain their fit for purpose. There are multiple generative AI models available today, each varying in capabilities, performance, cost, and specialties. Choosing the right model depends on an organization's overall strategy, the specific tasks at hand, and the budget allocated for AI initiatives. Alignment between the tool and the task is where AI delivers real, practical value rather than just being new or experimental. Governance practices must go beyond asking "how much does it cost?" and "what are the risks?"; they must also ask "why this tool for this task?" Understanding the unique strengths and limitations of different models and selecting one that best matches the intended outcome is critical for maximizing AI's benefits while managing costs and risks effectively.

Reference

1. Ian Goodfellow, Yoshua Bengio and Aaron Courville, Deep Learning (MIT Press, 2016).
2. Dan Milmo and Associated Press, 'Two US Lawyers Fined for Submitting Fake Court Citations from ChatGPT' The Guardian (23 June 2023) https://www.theguardian.com/technology/2023/jun/23/two-us-lawyers-fined-submitting-fake-court-citations-chatgp.
3. Eric Hilario et al, 'Generative AI for Pentesting: The Good, the Bad, the Ugly' (2024) 23 International Journal of Information Security https://doi.org/10.1007/s10207-024-00835-x.
4. Ibid.
5. Afia Sajeeda and B M Mainul Hossain, 'Exploring Generative Adversarial Networks and Adversarial Training' (2022) 3 International Journal of Cognitive Computing in Engineering.
6. Tushar Gupta et al, Generative AI with LangChain (Packt Publishing, 2023).
7. Vishwanatha Rao et al, 'Multimodal Generative AI for Medical Image Interpretation' (2025) Nature 639.
8. Stuart J Russell and Peter Norvig, Artificial Intelligence: A Modern Approach (4th ed, Pearson, 2020)
9. George F Luger, Artificial Intelligence: Structures and Strategies for Complex Problem Solving (6th ed, Pearson, 2009).
10. Ashish Vaswani et al, 'Attention Is All You Need' (2017) 31st Conference on Neural Information Processing Systems (NeurIPS).
11. Anthropic, Introducing Claude (2023) https://www.anthropic.com/news/introducing-claude
12. Generative AI for Legal Professionals: Top Use Cases (Thomson Reuters, 13 May 2025) https://legal.thomsonreuters.com/blog/generative-ai-for-legal-professionals-top-use-cases-tri/.
13. Lingjiao Chen, Matei Zaharia & James Zou, FrugalGPT: How to Use Large Language Models While Reducing Cost and Improving Performance (2023).
14. Eric Hilario et al, 'Generative AI for Pentesting: The Good, the Bad, the Ugly' (2024) 23 International Journal of Information Security.
15. Ibid
16. Sonam Singh, Amol Dhumane, Unmasking digital deceptions: An integrative review of deepfake detection, multimedia forensics, and cybersecurity challenges

(2025) MethodsX 15.
17. Andres Rodriguez, Deep Learning Systems: Algorithms, Compilers, and Processors for Large-Scale Production (Morgan & Claypool, 2021); Eric Hilario et al (2024)
18. DALL·E 2 (OpenAI) https://openai.com/index/dall-e-2/
19. Laurie Clarke, "When AI can make art – what does it mean for creativity? Dall-E, Midjourney" The Guardian (UK, 12 November 2022) https://www.theguardian.com/technology/2022/nov/12/when-ai-can-make-art-what-does-it-mean-for-creativity-dall-e-midjourney.
20. Zach Schor, "Andersen v. Stability AI: The Landmark Case Unpacking the Copyright Risks of AI Image Generators" (2 December 2024) NYU Journal of Intellectual Property & Entertainment Law (JIPEL Blog) https://jipel.law.nyu.edu/andersen-v-stability-ai-the-landmark-case-unpacking-the-copyright-risks-of-ai-image-generators/; Chloe Veltman, 'Anthropic Settles with Authors in First-of-Its-Kind AI Copyright Infringement Lawsuit' The New York Times (5 September 2025).
21. Patrick Lewis et al, "Retrieval-Augmented Generation for Knowledge-Intensive NLP Tasks" (2020) preprint https://arxiv.org/abs/2005.11401.
22. What Is Retrieval-Augmented Generation (RAG)? (Amazon Web Services) https://aws.amazon.com/what-is/retrieval-augmented-generation/.
23. Dario Amodei, 'The Urgency of Interpretability' (Blog post, April 2025) https://www.darioamodei.com/post/the-urgency-of-interpretability
24. Donna McGeorge, The Chat GPT Revolution: Get Curious. Get Productive and Get Creative with AI (2024, John Wiley & Sons Australia, Ltd.).
25. Ibid.
26. MicroAI, 'Lightweight AI – What it Means and Where it's Used' (2025) https://www.micro.ai/blog/lightweight-ai-what-it-means-and-where-its-used.
27. Office of the Victorian Information Commissioner, 'Artificial Intelligence and Privacy – Issues and Challenges' (2020) https://ovic.vic.gov.au/privacy/resources-for-organisations/artificial-intelligence-and-privacy-issues-and-challenges/.
28. Gemini Pro vs. GPT-3.5: How Google's New LLM Stacks Up (Akkio, 3 January 2024) https://www.akkio.com/post/gemini-pro-vs-gpt-3-5, OpenAI, "GPT-4o: OpenAI's First Multimodal Model" (OpenAI Blog, 13 May 2024) https://openai.com/index/hello-gpt-4o/.
29. Vishwanatha Rao et al, "Multimodal Generative AI for Medical Image Interpretation" (2025) Nature 639.

30. Antonio Mastropaolo, Luca Pascarella, Emanuela Guglielmi, Matteo Ciniselli, Simone Scalabrino, Rocco Oliveto and Gabriele Bavota, 'On the Robustness of Code Generation Techniques: An Empirical Study on GitHub Copilot' (Preprint, arXiv 2302.00438, February 2023) https://arxiv.org/abs/2302.00438.

"To Drive OR Not To Drive, that is the Prompt"

CHAPTER 6

Learning to Drive

When we learn to drive, we begin by understanding the gears, clutch, brake, accelerator, and steering wheel. Whether the car is manual or automatic, petrol or electric, every vehicle demands the same basics. Fuel for energy, precision to control speed, and judgment to brake at the right moment. Mastery comes from awareness, knowing how each control interacts with the other and when to adjust. Confidence is built not by speed, but by control.

The same principle applies to operating AI. Before taking action, we need to understand the controls that translate human intent into action. Prompts are those controls.[1] Tools like Copilot, ChatGPT, and Gemini respond directly to the quality of the instructions they are given. Knowing how to prompt them well is like learning to handle the steering wheel with precision. The goal is not just to use GenAI, but to drive it with skill, confidence, and purpose.

Taking the Wheel: Understanding Prompts

Very few people learn to drive by speeding down the highway. Most begin by understanding what happens in the driver's seat: the gears, pedals, and steering wheel that control the car. In the same way, every AI tool has controls that turn intention into motion. Before expecting precision from AI, we need to understand how to guide it. Prompts are the steering controls.

A prompt is the steering wheel that directs the system toward a result. It determines whether the experience is smooth and efficient or unpredictable and chaotic. Strong prompts help users achieve meaningful outcomes and reduce wasted energy. This keeps the model tuned and responsive instead of overworked.

A driver who misjudges the clutch or accelerates too quickly risks stalling or losing control. Generative AI is no different. The model is a powerful engine, but prompts are the steering mechanism.

When ChatGPT first appeared in late 2022, the world's attention was not only on the sophistication of its underlying model but on the human experience of control.

Prompt engineering emerged from this collective learning as a new kind of driving school. For the first time, real value came not just from the engine but from the craftsmanship of the person guiding it. Businesses quickly discovered that mastery of prompting could accelerate workflows, uncover insight, and reduce operational drag.

Prompting is not the same as searching. A search engine retrieves existing paths. Prompting, by contrast, is active driving. It requires the driver to frame intent, provide context, and decide how much control to apply. Poor prompts are like vague directions that lead nowhere. Strong prompts resemble precise routes programmed into a GPS, defining where to go and how to get there.

No matter how sophisticated or autonomous the vehicle, the quality of the journey depends on the driver. Every employee, manager, and customer who interacts with AI now sits behind the wheel. Their decisions determine speed, safety, and efficiency. When directions are careless, when confidential data is entered into public tools, or when outputs are trusted without verification, the whole system faces risk.

Driving without discipline can lead to collisions with regulatory, ethical, or operational hazards.[2] To make progress, drivers must stay alert, use proper signals, and treat prompts as the steering wheel of intelligence itself. Success depends not on how powerful the system is, but on how responsibly it is driven.

Using the Gears: What is a Prompt?

Every car, no matter how powerful, relies on a system that translates the driver's movements into action. The steering wheel turns the tires, the pedals regulate speed, and the gearbox determines how efficiently power is delivered. Together, they form the anatomy of the car. Generative AI operates in the same way. Generative AI works the same way. The prompt is its steering system. It is the link between human direction and machine motion. Understanding how this system functions is essential for using AI with accuracy, safety, and purpose.

A car may possess horsepower and design, yet the experience still depends on how the driver sets direction. Prompts play that same role. A vague or weak prompt is like saying "drive somewhere nice." The vehicle will move, but the path will be uncertain and the result disappointing. A clear and structured prompt[3], on the other hand, is like entering a precise address into a GPS. It defines a destination, outlines boundaries, and organizes the route ahead.

Strong prompts have key components that make the difference between a smooth drive and a stressful one. Understanding these elements helps drivers handle AI responsibly and get the best possible performance from the machine.

Setting the Scene: Role and Context

Every journey starts by knowing direction and conditions. Setting context for an AI model is the equivalent of deciding whether you are navigating through a city, driving a mountain pass, or cruising on a highway. The environment determines how you use the controls and how cautious or aggressive you should be. Providing context allows the model to filter information with accuracy and relevance.[4]

For example, a vague prompt such as "Write a summary of this report" leaves the model uncertain of tone or audience. A stronger version such as "Act as a financial analyst preparing a board briefing and summarize this report, highlighting regulatory implications" instructs the model where it is and how to drive. This steering context shapes everything that follows and ensures the route supports the destination you intend.[5]

Choosing the Route: Task and Instruction

Once the context is set, the next step is direction. In driving terms, the task is the destination. It is the GPS input that tells the system where to go. Without it, progression is aimless, costing time and resources. In prompting, the task must be clear, purposeful, and measurable.[6]

A prompt like "Write about data privacy" is the equivalent of driving until the road ends. The result may reach somewhere, but not where the driver hoped. A stronger instruction such as "Draft a 300-word overview of data privacy obligations under the Australian Privacy Principles suitable for inclusion in an internal training manual" specifies the route and helps the model stay aligned with the business purpose. The clearer the instruction, the smoother and more reliable the journey. [7]

Knowing the Limits: Constraints

Driving safely means respecting speed limits, lanes, and signals. Constraints in prompting play the same role. They define the rules of the road and prevent the system from veering into irrelevant or risky territory. Constraints manage both efficiency and safety by defining the scope of what the model should produce. [8]

A vague request like "Summarize this dataset" gives the engine freedom to roam aimlessly. A constrained prompt such as "Summarize this dataset in 250 words focusing only on year-over-year customer churn trends" provides clear rules and keeps the drive efficient. [9] Precision saves fuel, time, and cost. It ensures that no energy is wasted on unnecessary turns or confusion. The clearer the boundaries, the smoother and more predictable the trip.

Reading the Dashboard: Format and Output

The final aspect of the journey is how you receive information along the way. Some drivers prefer map views, others rely on voice directions. In the AI context, the format defines how results appear. The same destination can be reached in multiple ways, but the chosen display determines how quickly the driver understands and acts.

A prompt that says "Tell me about risks of AI adoption" opens a wide, unstructured highway. The model may deliver pages of winding narrative. A stronger prompt such as "Provide a table with three columns: risk category, board-level implication, and example mitigation" adds structure to the route and brings clarity to the arrival point. Well-defined formats transform raw responses into decision-ready output and make information immediately usable at business speed.[10]

Driving with Discipline

Understanding the anatomy of a prompt is about control and safety, not just technical skill. Large organizations rely on thousands of employees who, consciously or not, act as drivers of AI tools every day. Weak prompting wastes time, increases costs, and risks sending confidential data onto unsafe roads. Training staff to prompt with structure, precision, and restraint is the equivalent of teaching road rules before handing over the keys.

Prompt literacy has become as necessary as spreadsheet or presentation skills once were.Companies that invest in training will find their teams driving faster, more safely, and with lower operational costs. Those that neglect it will encounter traffic jams of inefficiency and collisions of misinformation.

Some predict that prompts will fade as AI grows more intuitive, shifting from written commands to natural speech or context-aware systems. Even if that happens, the underlying principle remains unchanged.[11] A responsible driver must know how to convey intent with clarity and judgment. Whether through text, voice, or gesture, the road between human thought and machine action still demands careful navigation.

Skilled prompting transforms raw power into purposeful direction. The The prompt framework brings order to that process, ensuring every interaction produces results that are accurate, safe, and useful. To support this, the following Prompting Dashboard serves as a quick reference. It outlines the key components, contrasts weak and strong prompts, and offers tips to guide AI with clarity and confidence.

Table 1: The Prompting Dashboard

Component	Weak Prompt	Strong Prompt	Tip for Better Prompting
Role and Context	"Write a summary of this report."	"Act as a financial analyst preparing a board briefing and summarize this report, highlighting regulatory implications."	Begin with who the model should be and what role or situation it's in.
Task and Instruction	"Write about data privacy."	"Draft a 300-word overview of data privacy obligations under the Australian Privacy Principles for an internal training manual."	Define what you want, for whom, and why.
Constraints	"Summarize this dataset."	"Summarize this dataset in 250 words focusing on year-over-year customer churn trends."	Use limits (e.g., word count or focus) to save time and energy.
Format and Output	"Tell me about risks of scams"	"Provide a table with three columns: risk category, implication, and controls."	Ask for lists, tables, or structured output for clarity.
Tone and Style	"Explain cybersecurity."	"Explain cybersecurity to a non-technical individual in a clear and confident tone, using simple and clear sentences."	Always state the tone and audience.
Verification	Accepts the output as correct.	Reviews and tests the result before using it.	Always review, verify, and ask for reasoning or sources.

Reference

1. IŞIN, Z. C., FİDAN, H., IŞIN, T., & TOPCU, M. K. Prompt Engineering 101 (Independently published, 2024) 10.
2. Ibid.
3. Donna McGeorge, The ChatGPT Revolution: Get Curious, Get Productive and Get Creative with AI (John Wiley & Sons Australia, 2024).
4. Clint Bodungen, ChatGPT for Cybersecurity Cookbook: Learn Practical Generative AI Recipes to Supercharge Your Cybersecurity Skills (Packt Publishing, 2024).
5. Donna McGeorge, The ChatGPT Revolution: Get Curious, Get Productive and Get Creative with AI (John Wiley & Sons Australia, 2024).
6. Harvard University Information Technology, Getting Started with Prompts for Text-Based Generative AI tool, (2023) https://www.huit.harvard.edu/news/ai-prompts.
7. Donna McGeorge, The ChatGPT Revolution: Get Curious, Get Productive and Get Creative with AI (John Wiley & Sons Australia, 2024).
8. Ibid.
9. IŞIN, Z. C., FİDAN, H., IŞIN, T., & TOPCU, M. K. Prompt Engineering 101 (Independently published, 2024) 10.
10. Gilbert Mizrahi, Unlocking the Secrets of Prompt Engineering: Master the Art of Creative Language Generation to Accelerate Your Journey from Novice to Pro Unlocking the Secrets of Prompt Engineering (Pakt, 2024).
11. Ibid.

Executive Summary (Part 1)

In this part, artificial intelligence was explored from its early rule-based origins to adaptive learning machines capable of recognising patterns, reasoning, and creating. Its evolution revealed how systems shifted from following instructions to improving through experience. A distinction was drawn between AI models and AI products: the model is the engine beneath the surface, while the product is the vehicle that carries its capability to users.

Key AI practices such as computer vision, natural language processing, and reasoning models were examined to show how data is translated into action across real-world settings. Learning methods including supervised, unsupervised, and reinforcement learning were introduced to explain how machines learn through instruction, discovery, or experience.

Building on this foundation, Generative AI was presented as an extension of machine learning. Large Language Models such as ChatGPT demonstrate how probability can be shaped into creativity, generating new text, images, and ideas rather than simply interpreting existing data. These systems show how creativity can emerge from structured learning, allowing machines to compose, converse, and assist at scale.

Prompting was then introduced as the human interface that guides these systems. Prompts act as the steering wheel of AI, translating intent into direction and ensuring that creativity remains controlled and purposeful.

Together, these elements show that AI is not a single technology but a connected system of engines, controls, and human judgment. Understanding how each component works and interacts with others is essential for responsible use.

KEY TAKE AWAYS

- There is a distinction between an AI system, an AI product, and an AI model. The three components play a different role in how intelligence is built.
- Every AI model is created through defined AI practices such as machine learning and deep learning, combining algorithms with training data to shape performance and reliability.
- Algorithm choice and data quality directly influence how accurately and ethically a model performs its task.
- Generative AI represents a class of Large Language Models, where outputs depend on the clarity and strength of the prompt guiding them.

PART 2

Risk and Security

“The question is not whether AI creates risk. It is whether you are prepared to manage it.”

CHAPTER 7

You Business Your Risk

The story of artificial intelligence is often told as an achievement of creativity. These are machines that can generate text, images, even music at the click of a button. What is often missed is how easily those same machines can create problems for the very organizations that deploy them.

In January 2023, CNET, one of the world's best-known technology publications, quietly began publishing finance articles written by AI.[1] The experiment was meant to illustrate efficiency and innovation. Instead, it backfired. Reporters discovered dozens of errors and passages that looked suspiciously like plagiarism. What was designed to demonstrate authority ended up undermining it.[2]

That is the paradox boards are facing. Artificial intelligence is powerful and versatile. It can structure emails, draft reports, generate marketing campaigns, and create images or video in seconds. For many organizations, these capabilities hold the promise of faster output and lower cost. But as the CNET episode shows, the very power that makes AI attractive also makes it dangerous. A single error can ripple into reputational, legal, and financial harm at extraordinary speed.

The temptation is to move quickly. Competitors are experimenting. Vendors pitch seamless integration. Staff already use free versions on their own devices. The feeling is that delay means irrelevance. Yet haste comes with a hidden cost: risk multiplied by scale. What starts as a clever experiment in a back office can, if ungoverned, become a systemic exposure once deployed enterprise-wide. Boards are not just passengers on this ride. They are responsible for the steering wheel, the brakes, and the guardrails.

Risk itself is not an unfamiliar terrain. Boards deal with it constantly, whether in finance, cybersecurity, or regulation. It is the possibility of an event that affects objectives, measured by its likelihood and consequence.[3] But AI reshapes the landscape. Likelihood is difficult to calculate when systems behave probabilistically and vendors have ability to change their models overnight. Consequence is harder to determine when a single flawed output can escalate across categories such as operational inefficiency, regulatory scrutiny, and reputational harm.[4] In AI, risks that once sat neatly in silos now arrive as bundles.

The challenge is not to invent new governance but to extend the discipline already known. Tools such as the risk matrix, remain essential. By plotting likelihood against consequence, boards can distinguish which exposures require immediate oversight and which can be monitored or accepted. But while these tools are familiar, AI introduces new dynamics that stretch their application.

This chapter bridges the gap between traditional risk governance and the emerging realities of artificial intelligence. It begins by revisiting how risk is defined and assessed, drawing on established frameworks such as ISO 31000 to anchor AI in familiar territory. It then applies those same principles to the digital frontier, showing how categories like legal, financial, operational, and reputational risk evolve in a world where algorithms shape decisions and information moves at machine speed.

AI is not a niche concern or a technical 'side' issue. It is an enterprise-wide exposure that touches every function, from strategy and culture to compliance and customer trust. Boards that overlook this connection will face risks they could have anticipated but failed to prepare for. Those that integrate AI into existing risk frameworks, define their appetite for exposure, and demand structured, transparent reporting will not only mitigate risk but also build the confidence to innovate responsibly.

Understanding Risk

Every organization face risk differently.[5] Supply chains can fail. Systems can be hacked. Lawsuits can emerge from contracts written years before. Directors are accustomed to scanning the horizon for threats, weighing them against opportunities, and deciding where the organization should take a calculated leap and where it should hold back. In that sense, artificial intelligence is not new ground. What is new is the shape of the risk: faster to emerge, harder to quantify, and more visible when it fails.

The key is to recognize that AI risks are not abstract. They are real, immediate, and capable of cascading across categories. A single mistake, an incorrect AI-generated policy, a flawed financial analysis, or a careless employee pasting confidential data into a chatbot, can trigger regulatory scrutiny, reputational backlash, and operational disruption in the same week. In a connected world, those events happen at speed.

In moments of technological excitement, it is tempting to treat risk as a secondary matter. The narrative of competitive advantage is seductive: "Our competitors are using AI to cut costs and we can't afford to be left behind."

The vendor story is equally alluring: "Integration is seamless, risk is minimal, productivity gains are immediate." And at the employee level, shadow adoption has already begun. Staff are experimenting with free AI tools outside of official channels.

Without a framework, boards are left reacting to a narrative. One week it is a story of an airline chatbot giving incorrect fare advice and forcing the company to honor a ticket.[6] The next, it is a university caught flat-footed as students use generative AI to write essays at scale. Or a global technology publication, CNET, which quietly experimented with AI-written finance articles only to be forced into public correction when the pieces were found to contain factual errors and plagiarized passages.[7] These are not science-fiction hypotheticals. They are live examples of what happens when AI is deployed without governance.

What boards need is not more hype, but structure. A way to cut through the noise and decide which risks deserve escalation, which can be accepted, and where resources should be focused.

The Risk Matrix

The simplest and most widely used tool is the risk matrix.[8] It measures exposure along two dimensions: likelihood, or how probable an incident is, and consequence, or how severe the impact would be.[9] Plotting these axes produces four outcomes. Risks with low likelihood and low consequence are monitored. Risks with high likelihood but low consequence are managed. Risks with low likelihood but high consequence are insured against or planned for. Risks with both high likelihood and high consequence are treated immediately and elevated to the board. The tool is deceptively simple yet powerful because it transforms uncertainty into prioritization.

Consider the risk of employees pasting personally identifiable information (PII) into a generative AI tool. The likelihood is high, especially without training or guardrails, and the consequence is severe: regulatory penalties, privacy breaches, and reputational harm. When combined, this places the risk at the top right of the matrix, representing extreme exposure that requires action. This is not an IT problem but a governance one.

Now consider the risk of a model generating biased job descriptions that unintentionally discriminate.[10] The likelihood may be moderate, with legal and reputational consequences. Plotted on the grid, it may not require the same urgency as a data breach, but it still demands mitigation through review and bias testing.[11] These examples show how a simple risk matrix helps organizations navigate which risks fall within appetite, which require treatment, and which cannot be accepted.

Why AI Complicates the Matrix

AI complicates the Risk Matrix for three reasons making it harder to place risks accurately. The first is probabilistic behavior.[12] Unlike traditional systems, AI models do not follow deterministic rules. They generate outputs based on statistical prediction. This means the same input can produce different outputs at different times. Estimating likelihood becomes difficult.

The second is the cascading consequences.[13] Traditional risks often fall neatly into one category such as financial loss, an operational delay, or a compliance breach. AI risks blur the boundaries. A chatbot hallucination might start as an operational issue, but if it reaches customers, it becomes reputational or even financial.

The third is diffusion of ownership. AI is used across the enterprise and through vendors, so risk is no longer contained within one function.[14] Marketing, HR, finance, and legal may all adopt AI tools, whether provided internally or by external providers. This diffusion makes accountability harder. A single employee can create exposure without realizing it. Boards must therefore weigh cultural and training risks alongside technical ones.

Tools like a risk matrix are useful, but they are not enough on their own. They give the board visibility, yet managing AI risk also requires responsible adoption across the organisation. This means ensuring employees are trained in safe use, vendors are assessed for transparency, and leaders set the tone that innovation is encouraged but boundaries are clear.

The ISO 31000 framework supports this approach through an ongoing cycle of identifying, assessing, treating, monitoring, and communicating risks.[15] It reminds organisations that risk management is not a one-time exercise. Risks must be reviewed and adjusted as models evolve and new uses emerge.

In the United States, NIST's AI Risk Management Framework complements this approach, emphasizing trustworthiness, transparency, and accountability. In Australia, APRA's CPS 234 on information security[16] and CPS 230 on operational resilience provide clear guidance. The details differ, but the principle is the same. AI risks must be assessed and treated with the same rigor as financial, legal, and operational risks. Frameworks alone, however, do not tell the whole story. Seeing risk on paper is one thing; watching it unfold in practice is another. The following examples illustrate how AI risks manifest across sectors and why ongoing review and control are essential.

Finance: A global bank pilots a generative AI chatbot to help customers with queries. Within days, it gives inaccurate credit advice. The consequence is only customer dissatisfaction but potential regulatory action. Likelihood is high, consequence severe.
Healthcare: A hospital uses AI to summarize patient notes. A small inaccuracy in a summary leads to a misdiagnosis. The operational risk is high, but so too are the reputational and legal consequences.
Education: Universities discover students using AI to draft essays. The likelihood is almost certain. The consequence is reputational damage if academic standards are compromised.
Government: An agency releases AI-drafted communication that contains errors. Public trust is eroded, political scrutiny follows. The consequence far outweighs the initial intent of efficiency.

These example illustrate that AI risks are not contained to one category. They cross over industry sectors, making the board's role in coordination and oversight essential.

Another layer boards must consider is risk appetite. Every organization has one, whether written down or not. If considering AI, risk appetite must be explicit. Without clear appetite, management is left guessing.

Residual risk is equally important.[17] No control is perfect. Even with strong training, employees will occasionally paste sensitive information into public systems. Even with human review, hallucinations will slip through. Boards should ask: given our controls, what exposure is left? Are we comfortable carrying it? [18]

Being explicit about appetite and residual risk ensures that when incidents occur, and they most likely will, that the organization is not blindsided. Instead, it can respond proportionately, knowing the exposure had been considered in advance.

Finally, decision makers must resist the temptation to treat AI risk as a matter for specialists. Of course, technical teams will handle implementation, compliance teams will manage controls, and risk teams will maintain registers. But the strategic questions should be directed towards appetite, prioritization, residual exposure, or vendor dependency. Factually, AI outputs are public. They carry your brand, your credibility, your trust. When they fail, the board, not just the CIO, is accountable.[19]

Every board is familiar with risk. A good risk register is like a map. It is a collection of threats, each named, described, and plotted with coordinates that make them easier to compare.[20] The form is deceptively simple: risk name, description, inherent risk (before controls), causes, impacts, controls, residual risk (after controls), and any additional treatments required. What matters is not the format itself but the discipline it enforces. It forces very different risks, from supply chain disruption to financial fraud to cyber intrusion and often at the same scale. It lets directors see, at a glance, which exposures can be accepted at a business unit level and which must be escalated to the boardroom.

That consistency is the bridge we need for AI. Generative systems may feel new, strange, even exotic. But the framework for assessing them is already embedded in businesses. The same template that tracks operational risk in logistics can be used to track prompt misuse in AI. The same logic that maps the impact of a data breach applies when a model hallucinates a customer's medical history. Risk assessment does not need reinvention. It needs adaption.

Let's look at a possible scenario involving a data breach. An employee uses an AI system to draft a client report and, for convenience, uploads invoice records or tax return details to improve the output. The risk is clear: exposure of confidential financial data through AI prompts. Sensitive information entered into public or poorly governed systems can lead to reputational, regulatory, and financial consequences.

The causes are varied. Human factors play a role when staff assume prompts are private or use AI tools without understanding where the data goes. Technology gaps contribute if there are no restrictions on file uploads or monitoring of outbound prompt traffic. Cultural pressures also matter when speed is valued over caution or when policies lag behind enthusiasm.

The impacts can be significant. Reputationally, customers and stakeholders may perceive the organization as careless with data. Regulators may find infringement with the Privacy Act, GDPR, or other sector-specific obligations. Financially, the costs can include fines, legal fees, remediation expenses, and the longer-term loss of trust and business.

Current controls are in place but not foolproof. End-user agreements help, but they rely on compliance. Information classification and sensitivity labels work, but they can be bypassed. Data Loss Prevention tools provide partial coverage, though not always for clipboard activity. Allow-listing sanctioned AI tools is effective for known platforms, while ongoing user training on prompt risks is valuable but highly dependent on culture.

The residual risk remains moderate. Existing measures reduce likelihood but cannot fully prevent staff from using external AI systems. Additional safeguards, such as endpoint clipboard monitoring or enterprise AI gateways, are recommended.

The value of a structured template is in the discipline it enforces. It makes new risks legible in the same language as old ones. Without that structure, AI risks remain anecdotes: a Samsung engineer pastes confidential code into ChatGPT, an airline chatbot misinforms a passenger, a newsroom publishes flawed AI-written articles. Each case may be alarming, but without a common frame they stay stories rather than strategy

With structure, the board can see these events as categories rather than curiosities. Incidents such as the Samsung, which exposed data through unsanctioned tools,[21] or the CNET episode that lost credibility through unchecked AI content, reveal different facets of risk. Each can then be plotted for likelihood, consequence, controls, and residual risk. The point is not to predict every possible incident but to give directors a way to sort the noise into a usable map.

It is tempting to see AI risk only through a cybersecurity lens, since breaches and data loss dominate the headlines. But AI touches every category of risk. Operationally, AI-generated code can introduce vulnerabilities. Reputationally, AI-written marketing content may plagiarize competitors. Regulators may view AI-driven decisions as discriminatory. Financially, vendor lock-in can drive escalating costs. And from a people perspective, over-reliance on AI can erode critical skills.

Each of these risks can be described, mapped, and compared using the same template. The board's role is not to master the mechanics but to ensure management applies the discipline consistently and reports results in a form directors can act on.

Inherent vs Residual Risk

One distinction worth underlining is between inherent risk and residual risk. Inherent risk is the level of exposure before controls are applied.[22] Residual risk is what remains after controls are in place. Boards should always ask: What is the residual risk, and are we comfortable carrying it?

In the context of AI, this distinction is important. As mentioned before, no control is perfect. For example, training reduces, but does not eliminate, the chance of employees pasting sensitive data. Human review reduces, but does not eliminate, the chance of hallucinated outputs reaching customers. Even with the best controls, residual exposure remains. The board's role is to decide whether that residual fits within the organization's appetite.

The true value of a structured risk assessment is comparability. AI may feel unfamiliar, but when its risks are expressed in the same format as fraud, compliance, or supply chain disruption, they become easier to evaluate. Once aligned in structure, AI risks can be weighed against existing business exposures to determine priority and impact. Comparability prevents AI from being treated as either a mystical threat or a trivial gadget. It helps directors see where AI fits within the organization's broader risk landscape and act with confidence rather than uncertainty.

This same structure prepares the board for what comes next. Once directors can view AI risks through familiar categories, they can examine the specific exploits that make those risks real: prompt injection, model bias, adversarial inputs, and deepfakes. Without structure, these threats seem overwhelming. With it, they can be assessed like any other risk such as naming the issue, understanding the cause, estimating the impact, and assigning controls.

AI may be complex, but governance already provides the tools to make it manageable. Directors only need to insist that emerging risks are presented in a clear and consistent format. That expectation turns uncertainty into oversight and enables boards to stay proactive instead of reactive.

Ultimately, managing AI risk is not about predicting every failure or eliminating uncertainty. It is about applying the same discipline that already underpins effective governance, including clarity, consistency, and accountability. When boards treat AI as part of the existing risk landscape rather than a separate domain, they strengthen resilience across the entire enterprise. The goal is not to control everything that the technology can do but rather to ensure the organization remains in control of what truly matters: trust, responsibility, and purpose.

Reference

1. Emily Bell, 'A Fake News Frenzy: Why ChatGPT Could Be Disastrous for Truth in Journalism' (The Guardian, 3 March 2023) https://www.theguardian.com/commentisfree/2023/mar/03/fake-news-chatgpt-truth-journalism-disinformation.
2. Ibid.
3. G Russell Hubbard, Strategic Management: Thinking, Analysis and Action (Pearson Education, 2000).
4. K Scott Griffith, The Leader's Guide to Managing Risk: A Proven Method to Build Resilience and Reliability (Harvard Business Review Press, 2014).
5. Gregory J Falco and Eric Rosenbach, Confronting Cyber Risk: An Embedded Endurance Strategy for Cybersecurity (Oxford University Press, 2021).
6. Maria Yagoda, 'Airline Held Liable for Its Chatbot Giving Passenger Bad Advice – What This Means for Travelers' (BBC Travel, 22 February 2024) https://www.bbc.com/travel/article/20240222-air-canada-chatbot-misinformation-what-travellers-should-know.
7. Emily Bell, 'A Fake News Frenzy: Why ChatGPT Could Be Disastrous for Truth in Journalism' (The Guardian, 3 March 2023) https://www.theguardian.com/commentisfree/2023/mar/03/fake-news-chatgpt-truth-journalism-disinformation.
8. Brian Knutson and Scott A Huettel, 'The Risk Matrix' (2015) 5 Current Opinion in Behavioral Sciences.
9. Ibid.
10. Paul R Daugherty and H James Wilson, Radically Human: How New Technology Is Transforming Business and Shaping Our Future (Harvard Business Review Press, 2022).
11. Douglas W Hubbard and Richard Seiersen, How to Measure Anything in Cybersecurity Risk (John Wiley & Sons, 2016).
12. Stuart J Russell and Peter Norvig, Artificial Intelligence: A Modern Approach (4th ed, Pearson, 2020).
13. K Scott Griffith, The Leader's Guide to Managing Risk (Harvard Business Review Press, 2014).
14. Gregory J Falco and Eric Rosenbach, Confronting Cyber Risk: An Embedded Endurance Strategy for Cybersecurity (Oxford University Press, 2021).
15. International Organization for Standardization, ISO 31000:2018 – Risk Management – Guidelines (ISO, 2018) https://www.iso.org/standard/65694.html.

16. Australian Prudential Regulation Authority, Prudential Standard CPS 234 – Information Security (July 2019) https://www.apra.gov.au/sites/default/files/cps_234_july_2019_for_public_release.pdf.
17. International Organization for Standardization, ISO 31000:2018 – Risk Management – Guidelines (ISO, 2018) https://www.iso.org/standard/65694.html.
18. K Scott Griffith, The Leader's Guide to Managing Risk (Harvard Business Review Press, 2014).
19. Corporations Act 2001 (Cth) ss 180–183.
20. Gregory J Falco and Eric Rosenbach, Confronting Cyber Risk: An Embedded Endurance Strategy for Cybersecurity (Oxford University Press, 2022).
21. Davey Winder, 'Infostealer Strikes Samsung: 270,000 Records Stolen' Forbes (31 March 2025) https://www.forbes.com/sites/daveywinder/2025/03/31/infostealer-strikes-samsung-270000-records-stolen/.
22. ISO, ISO 31000:2018 Risk Management – Guidelines (International Organization for Standardization, 2018) https://www.iso.org/standard/65694.html.

"AI risk is not science fiction. It is business exposure, multiplied by speed and scale."

CHAPTER 8

Exploiting AI

Artificial intelligence, arguably, began with a single question: "Can a machine have human intelligence?"[1] Since its launch, AI was often used more for entertainment than for anything of substance. Early outputs were harmless, whimsical, and easy to dismiss. That changed when businesses began to see its potential value. Soon, AI became the new shiny tool. Like most tools of this magnitude, in the wrong hands, it can be weaponised. Today, AI is not just entertainment. It is an instrument of exploitation.

The adoption and misuse have been immediate. Scammers now use AI-generated voices to impersonate family members in urgent calls for money.[2] In politics, deepfake videos distort words that were never said.[3] In recruitment, synthetic résumés and cloned LinkedIn profiles flood hiring platforms. Even in private life, AI-generated romantic partners are deployed to manipulate, harvest data, or extort.[4] None of this is speculative. It is happening now and at scale.

The common factor is not that machines are malicious, but that humans have recognised that these machines are capable of deception at industrial speed. A single fraudster can clone hundreds of voices in a day. A state actor can flood social media with fabricated personas. A disgruntled insider can use AI to generate convincing documents to discredit leadership. Where once deception required skill, time, and resources, it now requires little more than a prompt.

From a governance perspective, this creates a significant challenge. Manipulation has become cheap and abundant. The barrier to entry has collapsed, while the potential for harm has multiplied. The result is a completely altered risk environment. Boards can no longer afford to assume that AI-related exploits will be caught by a security team. These emerging risks expose organisations to vulnerabilities that are difficult to isolate, measure, or predict.

A single deepfake can spark reputational damage. An AI-generated phishing email can bypass staff training that worked against traditional scams.[4] A synthetic résumé can introduce an insider threat. Each of these risks falls not in the abstract world of emerging technology but in the core domains boards already oversee: trust, compliance, financial integrity, and brand.

One of the most persistent misconceptions is that these risks come from the machines themselves. Some argue that AI is plotting against its users. It is not. AI has no intent. The threat lies in how humans use it. The machine does not scheme, but it executes schemes — and it does so using our faces, our voices, and our data. Because the output is often convincing, the victims are not only systems but people. Customers are misled, employees are tricked, and investors are spooked.

AI is as powerful as it is deceptive because of its accessibility. A model designed to draft marketing copy can just as easily generate a phishing campaign. A system that creates product designs can also fabricate evidence. A chatbot that answers customer questions can persuade an employee to hand over credentials. The same capability that makes AI attractive to business also makes it attractive to bad actors.

These are emerging risks. They are already reshaping finance, security, employment, and personal relationships. Boards that fail to anticipate them will find themselves reacting to crises instead of guiding outcomes.

A proactive board will recognise that every AI risk must be paired with a control. Understanding this relationship is key to reading any AI risk assessment. When directors can see both the risk and its mitigant, they can ask sharper questions, identify gaps, and challenge whether controls are sufficient. They can also ensure that the organisation's risk appetite is defined and applied consistently across all AI systems.

This chapter introduces several categories of AI risk. Each is explained in terms of its potential impact on the organisation, followed by practical ways to mitigate it. The controls illustrate how structured oversight can strengthen governance and reduce exposure.

DeepFakes

Imagine this scenario. A Fortune 500 CEO appears on screen during a livestream, announcing the sale of several companies.[5] Markets react immediately, journalists report, and employees scramble. Hours later the truth emerges. The video was fabricated by AI. The real CEO never spoke, yet stock prices shifted, competitors responded, and trust was damaged in ways no press release could ever repair.

That is the power of deepfakes. They are not crude edits but sophisticated simulations of sight and sound, producing video and voice content that appear authentic to the human eye and ear.[6] What once required a Hollywood studio can now be done on a laptop, using material drawn from LinkedIn headshots, YouTube panels, keynote speeches, podcasts, or even casual social posts.

Deepfakes undermine the most basic currency of business: trust. A board member joining a video meeting needs to know who is on the other side.[7] A bank conducting remote KYC (Know Your Customer) checks must be confident in identity. Regulators reviewing compliance evidence must be certain that testimony is genuine.

A model trained on thousands of images or recordings can mimic expressions, intonations, and micro-movements, generating real-time streams that adapt to lighting, head movement, and lip synchronization. Detection is difficult. Liveness checks such as blinking or head turns can be mimicked. Voice authentication can be bypassed with minutes of training. Even advanced biometric tools struggle to detect absence, not deception.

The exploit surface is already wide. In finance, fraudsters bypass KYC to open accounts under stolen identities.[8] In politics, fabricated speeches erode trust in institutions. In personal life, voice clones impersonate family members in urgent scam calls. A 2,137 percent rise in deepfake fraud attempts over three years shows how fast novelty has become systemic threat.[9]

The implications are strategic. A convincing fake can damage brand integrity overnight. Regulators may penalise firms if synthetic identities slip through. Financially, fraud losses and litigation costs can run into millions. For attackers, deepfakes are cheap to create; for defenders, authenticity is expensive to prove. This asymmetry is why regulators, including the U.S. Department of Homeland Security, now classify deepfakes as an emergent threat.[10]

This risk cannot be eliminated. Detection tools are in an arms race with adversaries, and residual exposure will remain. The machine does not scheme, but it executes schemes with your face, your voice, your data. That inversion is what makes deepfakes so dangerous.

Deepfakes are not just a cybersecurity problem. They are a trust problem. And trust is the foundation of markets, regulation, and reputation. Their danger in not in the level technical capability but in their ability to hijack sight, sound, and speech which we instinctively rely.

The machine does not scheme, but it executes schemes by using your face, your voice, your data. That inversion is what makes deepfakes so powerful, and why boards cannot afford to look away.

What does this mean in practice? Vendor contracts must shift liability and require up-to-date detection tools. Verification should go beyond static images to include liveness checks. Staff need training to recognize anomalies, with clear escalation paths when suspicion arises. Incident response and communication plans must also be tested, because when a deepfake hits, the damage spreads faster than the fact-check. Let's now turn to the controls that can help mitigate deepfake risk.

Contracts: Most organizations will rely on third parties for KYC, interview platforms, or communication tools. Contracts should transfer liability, require up-to-date detection capabilities, and define clear service levels for incident handling.

Verification: A single static image or video is no longer enough. Stronger processes can include liveness checks by turning the head, opening the mouth, or repeating randomized phrases. This make it harder for synthetic media to pass undetected.

Training and Awareness: Staff should understand that sight and sound are no longer reliable. Training should focus on identifying anomalies and escalating quickly if suspicion arises.

Process controls: Verification should never rest on one input. Entities should design layered decision points, such as pairing video verification with document checks, or requiring logins through a secure, centralized platform like Teams. Centralization provides consistency, monitoring, and audit trails

Human-in-the-Loop(HiTL): Not every check can or should be automated. In high-risk situations, there should be a human in the loop. This could be a face-to-face verification at a post office, bank, or authorized provider. While it adds an extra step in the process, it can add build trust in a quickly evolving world.

Access and monitoring: Multi-factor authentication provides an additional barrier, making it harder for attackers to succeed with only a deepfake. Network monitoring adds another layer, flagging unusual locations, browser fingerprints, or traffic patterns. In some cases, the data flow of a deepfake session may even differ from a normal call, creating red flags that automated systems can detect.

Voice Biometrics

In 2024, a Hong Kong company lost $40 million to a voice-cloning scam. A finance executive believed he was following the instructions of his CEO.[11] He wired funds to criminals who had convincingly imitated the executive's voice. The execution and size of financial loss made headlines. Interestingly, the tactic itself is not new. Five years earlier, in 2019, a U.K. energy executive had transferred $243,000 USD after receiving, what he thought, was a routine phone call from his German CEO.[12] Investigators later confirmed that the "CEO" was nothing more than a synthetic voice but persuasive enough to cause human error.

Our voice has been treated as proof of identity. Court testimony, verbal contracts, boardroom negotiations, all depend on the assumption that if you recognize the voice, you can trust the speaker. AI has fractured this fundamental methodology.

In many cases, individuals don't even know that their voice has been stored. The polite disclaimer, "This call may be recorded for training purposes," is often the only notice. Once captured, those recordings may be retained indefinitely, sometimes in poorly secured systems. When breached, sold, or traded, they become raw material for exploitation. The threat doesn't depend on what you post publicly. It may already exist in someone else's archive.

One no longer needs advanced labs or insider access to exploit voice biometrics. All that is required is access to voice data, which is, ironically, easy to find. With the rise of data breaches, it is likely that vast libraries of captured voices already sit on the dark web. Customer service calls, compliance recordings, sales surveys, and helpline logs are often kept for years. A single breach of those archives can expose thousands of hours of speech.

Unlike early text-to-speech systems, today's models can replicate pitch,cadence, and rhythm with uncanny accuracy.[13] Some can generate speech in real time, modulating tone to sound urgent, authoritative, or friendly depending on the scam. A voice clone is not just a static copy. It is a dynamic instrument, capable of improvising on the fly.

The problem becomes worse when voice biometrics are used for authentication.[14] Banks, insurers, and service providers increasingly use voice-prints to verify identity, analyzing unique vocal patterns like fingerprints. Customers are told, "Your voice is your password." But what happens when that voice can be reconstructed from leaked recordings? A system designed to strengthen security becomes a point of vulnerability. In many cases, individuals don't even know their voice has been stored.

Voice cloning has moved far beyond isolated incidents of fraud. It is now a tool for large-scale manipulation. In corporate settings, criminals impersonate executives to authorise payments or disclose sensitive information, exploiting the trust placed in familiar voices. In personal settings, families are targeted with cloned voices of children or parents pleading for help. In customer service, synthetic voices bypass authentication systems that rely on voiceprints. And in the public sphere, fabricated audio of politicians or executives can spread faster than any correction, damaging reputations before truth can intervene.

The imbalance between attacker and defender is striking. A cloned voice can be created in minutes and replicated endlessly. Verifying authenticity, by contrast, requires far greater time, expertise, and cost. Unlike video deepfakes, synthetic voices are harder to detect. People instinctively trust what they hear, and that trust becomes the point of exploitation.

The risk extends beyond individual incidents. It crosses financial, legal, operational, and reputational domains. Collecting or storing voice data without consent exposes organisations to regulatory action under privacy laws such as the GDPR and the Australian Privacy Act.[15] A breach involving biometric data can erode customer and investor confidence as quickly as any cyber incident.

Operational response is also affected. When every urgent call might be a deception, decision-making slows. If voice remains an approved channel for crisis management, the vulnerability deepens. Employees hesitate to act on instructions, customers question authenticity, and markets react to false statements before verification is possible. Once trust in spoken communication is lost, rebuilding it is slow, expensive, and uncertain.

Acknowledging the risk is only the beginning. Voice cloning cannot be eliminated, but its impact can be reduced through layered defences. These include contractual obligations for vendors handling audio data, technical verification tools that detect synthetic speech, and procedural safeguards that require secondary confirmation for voice-based instructions. Boards should ensure these controls are in place and regularly tested.

Let's now examine several practical controls that can help organisations mitigate this growing risk.eview several controls that can help mitigate this risk.

Criticality Classification: Classify activities into tiers of criticality. Routine inquiries may be authenticated by voice alone, but high-value or high-risk actions should require additional checks. This keeps friction proportionate while protecting the most sensitive processes.
Role and Social Media Policies: Executives and senior leaders are often profiled by attackers. Public sources such as LinkedIn, company websites, and conference bios can reveal who approves payments, who signs contracts, or who makes strategic decisions. That information becomes key for cyber criminals. Boards should set clear policies that limit how much detail staff share about their responsibilities. Keeping role descriptions broad reduces the intelligence criminals can exploit.

Voice-Clone Detection Tools: Technology can assist by analyzing calls for synthetic patterns. Voice-clone detection tools examine cadence, frequency, and anomalies that humans may not hear. While imperfect, they add an additional layer of defense when integrated into call-center and executive communication systems.

Liveness and Randomized Checks: Simple challenges, such as asking the caller to repeat a random phrase, answer an unscripted question, or count backwards, could expose a cloned voice. These checks disrupt the smooth playback of a synthetic stream and create friction for attackers.

Dual Authorization for Critical Actions: High-value decisions such as financial transfers, contract approvals, or major operational changes should never rest on a single voice instruction. Establishing thresholds where two approvers must confirm the action adds resilience. By requiring confirmation through separate channels, cloned voices lose their power as a single point of failure.

Multi-Factor Verification: Confidential actions should not rest solely on a single voice-based request. Supplementing with a second factor, such as a secure push notification, login to a corporate portal, or encrypted written confirmation, which creates an additional hurdle. Even if a voice is cloned, the instruction cannot proceed without further authentication.

Secondary Validation Channels: Critical voice instructions should always be verified through an independent channel. A quick Teams message, secure email, or in-person check ensures human judgment remains in the loop. This control raises the cost and complexity for attackers, forcing them to replicate more than just the voice.

Monitoring and Metadata Analysis: Beyond content, look at the call itself. Abnormal call origins, unusual traffic volumes, or metadata mismatches can signal suspicious activity. Centralized telephony environments can flag anomalies that human staff might miss in real time.

Synthetic personas and bots

In 2020, reporters traced a network of online accounts that looked like everyday users.[16] Their profile pictures showed friendly, photo-realistic faces. Their biographies described plausible jobs and hobbies. They shared articles, joined debates, and supported one another's posts. On the surface, they seemed like a diverse set of individuals. In reality, not one of them existed. The faces were generated by AI.[17] The posts were written by large language models. The network was coordinated to sway opinion in political debates.

What once looked like simple spam bots has transformed into something far more convincing. AI tools assist criminals create personas that are almost indistinguishable from real people. GANs and diffusion systems produce faces of people who never lived. Language models can craft biographies, posts, and interactions tailored to particular audiences. While automated frameworks link it all together by distributing content across hundreds of accounts in real time. The result is a synthetic community with a web of voices that appear independent but are, in fact, orchestrated.

Synthetic personas can distort reputation by creating the illusion of consensus around negative reviews or false claims. They can impersonate a company's staff, confuse customers and erode trust. They can overwhelm customer service systems with fabricated complaints, skew analytics and exhaust front-line teams. They can be used to manipulate sentiment, creating hype or panic within financial markets. And in compliance, they can slip through authentication checks, enabling fraud or money laundering under the cover of a fabricated identity.

What is important is that a single person can now generate and manage hundreds of convincing digital personas using off-the-shelf tools. Cloud computing makes the resources available by the hour and open-source toolkits lower the barrier to entry further.

That makes controls essential. It is not enough to acknowledge that synthetic personas exist. Organizations must ensure that concrete measures are in place to detect, verify, and respond. While bots and fake accounts are multiplying, so too are the tools and processes to counter them. What matters is selecting controls that are proportional to the exposure and embedding them into governance routines. Let us review several controls that can mitigate this risk.

Bot Detection and Monitoring: Invest in monitoring tools that analyze traffic patterns and behaviors, not just content. Synthetic personas often operate in coordinated bursts by posting at unusual times, repeating identical phrases, or amplifying each other's messages. Monitoring for these anomalies will help identify and take-down false narratives.
Multi-Channel Verification: A single channel is vulnerable to manipulation. Companies should publish critical updates across multiple official outlets such as the website, email bulletins, social media or investor portals. Equally important, customers and stakeholders should be encouraged to contact the company directly if they suspect misinformation. Clear instructions on where and how to verify information, may it be via a hot-line, a trusted email address, or a secure portal should be made know.
Employee and Customer Awareness: Staff in marketing, customer service, and investor relations must be trained to recognize signals of coordinated inauthentic behavior. Customers should also be educated on where to find "official" communication so that bots cannot easily impersonate the company or distort its message.
External Monitoring Services: Subscribing to third-party media monitoring or disinformation tracking services gives early warning when synthetic personas or bots target your organization. Alerts allow

leadership to respond quickly, issue clarifications, and engage regulators or platforms before reputational damage spreads.

Shadow AI

Not every exploit comes from hostile outsiders. Some of them come from inside the organization itself. A manager pastes bullet points into ChatGPT to draft a client message. A recruiter uses an online generator to polish a job ad. A board member asks Copilot to refine a strategy memo. None of these actions feels malicious. In fact, they seem helpful. This is an invisible layer of technology that operates outside of enterprise oversight. This is Shadow AI: the natural evolution of Shadow IT.

Most people have been exposed to AI as a supercharged search engine. It delivers quick results, making it even more tempting to use. Because success is immediate, they recommend it to colleagues. Within weeks, entire teams may be leaning on unsanctioned tools without IT's knowledge.

The convenience is obvious, but the risks are invisible. Every prompt, every document, every data type shared with an external system becomes potential exposure. Without logs, without audit trails, and without guardrails, this place the organize at regulatory, financial and reputational risk.

Shadow AI is becoming a growing concern with 98% of employees admitting to using unsanctioned apps, including shadow AI.[18] Shadow AI is not the behavior of a few careless individuals. It is a systemic reality. Nearly every organization faces this risk whether leadership acknowledges it or not.

One of the most common and most dangerous assumptions is that prompts remain private.[19] The reality is more complicated. Consumer tools such as ChatGPT Free or Plus, Gemini, and Perplexity often retain prompts. Some vendors use them for model training by default unless the user opts out. Others keep them for monitoring, debugging, or abuse prevention.

Enterprise tools such as ChatGPT Enterprise, Microsoft Copilot, or Gemini Enterprise offer stronger protections. OpenAI, for example, states that enterprise and API data is not used to train its models and is only retained briefly for abuse monitoring. Microsoft emphasizes that Copilot in M365 keeps data within the customer tenant, protected by the same compliance framework as other Office documents.

Unless your organization has a signed enterprise agreement with explicit guarantees, you cannot assume prompts are private. Once pasted into a consumer-grade service, sensitive data may be retained, logged, or analyzed, even if the vendor claims it is not used for training. Most large models capture prompts through telemetry for quality assurance, content filtering, and abuse detection. Even anonymized, this data can persist in logs or backups for extended periods and may be accessed by the vendor's personnel or third-party service providers.

Much of this exposure stems not from malice but from curiosity. Employees turn to public AI tools to speed up work or solve problems creatively, assuming the data they enter disappears once the session ends. In reality, those prompts can remain in vendor systems, beyond the organization's control.

Employees experimenting with free tools are effectively sending organizational data into systems governed by consumer terms of service. Only enterprise deployments provide enforceable guarantees, and even those require proper configuration to be effective.

According to Aditya Patel, Shadow AI is "your IT team's worst nightmare."[20] While monitoring tools can detect data in emails or file transfers, they can easily detect sensitive data that are exfiltrated through an AI prompt. To the system, ShadowAI is similar to web traffic. Data such as contracts, medical records, or financial forecast can leave slip through the network without a trace.

The problem is compounded when a data breach raises questions on how confidential data ended up in a public dataset with no audit trail. It is impossible tract who shared this data or which system was used. This lack of accountable place the board in a precarious position.

Shadow AI is not just a technical risk. It is a governance gap. Without visibility, accountability collapses. Without accountability, policy cannot be enforced. And without enforceable policy, the board cannot demonstrate to regulators, investors, or customers that AI use is under control.

This is why embedding AI into secure environments matters. Enterprise integrations like Microsoft Copilot, when configured correctly, allow organizations to apply existing controls such as classification, retention policies, sensitivity labels, data loss prevention and access restrictions. These are not perfect solutions. But they are safer than unmonitored use of consumer-grade tools. The choice is not between AI and no AI. It is between AI in the light and AI in the shadows.

When it comes to Shadow AI, the risk lies in what cannot be seen. And if it cannot be seen, it becomes harder to govern. Without governance, the organization may find itself accountable for risks it never anticipated. The following section outlines key controls that boards should expect to see in place.

Enterprise Integration: Using approved platforms like Microsoft Copilot, ChatGPT Enterprise, or Gemini Enterprise is one of the strongest safeguards available. These services keep prompts within the organization's own tenant and include contractual guarantees that data will not be reused for training. However, the protection is only as strong as the environment it operates in. If an organization lacks basic controls, such as data loss prevention, information labeling, or identity management through Entra ID, then adding AI may introduce the same risks as Shadow AI. The real difference between safe and unsafe adoption is not the software license, but the strength of the security framework supporting it.

Identity and Access Management: An organization can strengthen authentication with Entra ID, MFA (Multi-Factor Authentication), and conditional access. This ensures that only authorized users can connect to enterprise AI tools, and usage can be logged against each approved corporate user.

Whitelisting: Whitelisting can restricted access to unsanctioned AI sites. Network-level controls can block common consumer tools, ensuring staff use only approved systems with contractual safeguards.

Data Classification and Labeling: Data classification and labels can be applied to unstructured data to reduce the risk of employees uploading sensitive documents into an AI prompt. When these labels are tied to access restrictions and monitoring, they act as a visible reminder to staff and a technical safeguard that prevents mishandling of critical information.

Monitoring and Logging: Deploy tools that can detect unusual traffic patterns, such as large data uploads to endpoints, especially those related to AI. Logging and alerting provides traceability.

Training and Culture: Employees need to know that prompts are not private. Clear training and practical examples can shift behavior from convenience-driven to risk-aware.

Hallucinations

Artificial intelligence does not lie, at least not in the way humans do. It has no intent, no agenda, no motive to deceive. What it does instead is hallucinates. In the world of AI, a "hallucination" is a response that looks plausible, sounds authoritative, and is often delivered with confidence. The output, is in fact, completely fabricated. These are not a formatting glitch or typo. They are false legal citations, fabricated policies, misleading financial summaries, or fictitious medical advice.

The risk is in how confident the output is delivered. AI does not hesitate. It asserts. A fabricated case law is presented with a citation and docket number. A non-existent medical study is summarized in professional prose. The delivery is polished enough to disarm skepticism, especially for non-experts. From a psychological perspective, it resembles a pathological liar. [21]

Hallucinations happen because of how these systems are built.[22] A generative model is not a library that can look up an answer. It predicts the next word based on patterns in the data it has seen before.[23] When the data is missing or limited, it makes something up to fill the gap. When a question can be read in different ways, the model picks one and runs with it, even if it is wrong. And when asked to keep writing without a clear answer, the model will not admit it does not know. It will keep generating, even if the result is nonsense.

This is not a software bug that can be patched. It is part of how these systems work. Some tools try to reduce hallucinations by adding guardrails or connecting the model to external sources. These steps help, but they do not remove the problem. No system available today is immune.

In 2023, a lawyer in the United States filed a court brief that included several case references. On the surface, everything looked professional.[24] The format was correct, the citations looked real, and the writing was clear. The problem was that every case was fictitious. The lawyer only discovered this when the judge checked and asked for proof. The shock was not just that the cases were invented, but that the output looked so normal that even an experienced professional trusted it.

Hallucinations create a serious challenge for organizations. AI can generate text that looks true, even when it is not.[25] How do you verify the accuracy of information produced by a system when you have no visibility into its training data or reasoning? Traditional assurance depends on traceability, the ability to follow a statement back to its source. With generative AI that trail does not exist. The output is generated through billions of parameters, but the reasoning behind it cannot be inspected.

This lack of transparency creates two exposures: accountability and compliance. If a report contains fabricated data, who is responsible? Is it the employee who used the tool, the vendor who built it, or the board that authorized its use? Regulators will not accept "the AI did it" as a defense. Misleading outputs in financial reports, legal filings, or customer communications remain the organization's responsibility.

Hallucinations cannot be eliminated, but they can be managed if organizations foster a climate of skepticism and verification. Just as "trust but verify" became the mantra of cybersecurity, "generate but verify" must become the mantra of AI adoption.

AI will not lie to you, but it will not tell you the truth either. It will generate. The responsibility to verify lies with the organization. That responsibility cannot be outsourced to the vendor, delegated to employees without training, or postponed until after reputational damage is done. Let's review several mitigant to this risk.

Vendor Contracts and Audit Rights: AI vendors should provide transparency about confidence scoring, error rates, and testing. Contracts can require periodic audit reports, allowing organizations to validate whether the system meets agreed standards. This shifts oversight from blind reliance on a black box to measurable performance that can be independently checked.

Fact-checking against trusted sources: Every critical output should be verified against authoritative data. Financial numbers against internal ledgers, legal citations against official case law, etc. This prevents fabricated content from slipping into official use.

Subject Matter Experts (SMEs): Have domain experts review AI-generated content before it leaves the organization. Lawyers, compliance officers, or clinicians can spot errors that look plausible to non-experts. Their oversight is a strong line of defense.

Structured Review Process: Entities should require draft outputs to pass through approval steps before publication, filing, or client delivery. This shifts AI from "decision-maker" to "assistant" and catches errors early.

Disclosure Policies: Make it standard practice to note when AI has been used in creating reports, briefs, or communications. This builds transparency and avoids surprises if errors are uncovered later.

Training and Culture: Staff should be trained to treat AI outputs as drafts, not truths. Encourage a culture of skepticism where questioning results is expected, not penalized.

Monitoring and Metrics: Track how often AI outputs require correction. This gives boards visibility into reliability and highlights areas where additional safeguards are needed.

Model Drift

A model that once delivered reliable results, built on historical data, can slowly lose touch with reality as the world shifts. This is model drift. Accuracy and reliability decline when the data coming in no longer resembles the data the system was trained on.[26]

Drift is not dramatic like a system outage or a coding failure. It is quiet. The model keeps running. Dashboards keep reporting results. Confidence scores still look strong. Yet errors build up out of sight until they show up as financial loss, regulatory pressure, or damage to reputation.

The risk is subtle, but it is systemic. Take an e-commerce platform that relies on AI to detect fraud. At first, the system works well, flagging large, unusual transactions that stand out from typical consumer behavior. Over time, criminals adapt. They move to smaller, more frequent purchases that look ordinary. The fraud continues, but the model no longer sees it. Losses rise, not because the model crashed, but because it never learned the new tricks. This is drift in action. The environment changed. The model did not.

Drift is inevitable because environments evolve faster than static training data.[27] Consumer tastes can change overnight: a music service trained on last year's playlists may still push pop while audiences move to retro or jazz. Fraud tactics adapt too, with criminals spreading charges across small items to avoid detection. External shocks reshape behavior instantly, as the pandemic altered work and commuting, while events like Brexit or trade sanctions disrupted supply chains. Even in finance, models built on pre-inflation data may misjudge risk as rising rates make old spending patterns unreliable.

Drift comes in two forms. The first is data drift, where inputs change. For example, as a company expands internationally, addresses, currencies, or tax codes appear in formats the model has never seen before.[28]

The second is concept drift, where the link between input and outcome shifts.[29] Shopping behavior that once signaled loyalty may now indicate financial strain. In both cases, the model keeps solving yesterday's problem with outdated assumptions. It continues to run, but its accuracy erodes until mistakes surface as losses, compliance failures, or reputational harm.

The risk lies in the invisibility of drift. To boards, systems may appear to function normally, producing confident outputs even as accuracy slips and bias returns. By the time errors are obvious, costs can be high: loans denied unfairly, supply chains misaligned, diagnoses delayed, or financial risks mis-priced.

Drift is not a technical quirk. It is a structural risk that undermines trust, compliance, and financial performance if left unchecked. The responsibility is not to fine-tune models but to ensure robust monitoring, escalation, and accountability frameworks. Regulators in finance, healthcare, and insurance already demand evidence of these capabilities. Organizations unable to demonstrate drift detection and response risk being accused of systemic governance failure. With that in mind, we turn to the practical controls that can mitigate this risk.

Performance Monitoring: Drift often goes unnoticed because models keep running with apparent confidence. Continuous monitoring of input data, output accuracy, and fairness metrics helps detect when the model begins to diverge from expectations. Dashboards should track not just accuracy but also error rates, bias indicators, and trends over time.

Regular Retraining: No model remains accurate forever. Scheduled retraining with fresh, representative data is one of the most effective safeguards. This requires clear policies on how often retraining occurs, how new data is validated, and who signs off before deployment. Without a structured cycle, drift quietly accumulates.

Human-in-the-Loop Review: Automated decisions should not be fully trusted in dynamic environments. Periodic sampling and human review of outputs provide a check against unnoticed errors. Subject matter experts can spot subtle mis-classifications or biases that metrics alone might miss, ensuring models remain aligned with business realities.

Scenario and Stress Testing: External shocks such as pandemics, market shifts, or regulatory changes can overturn assumptions overnight. Scenario testing helps organizations model how AI systems respond to new conditions. Boards should expect management to show how models would perform if customer behavior, fraud patterns, or economic indicators changed suddenly.

Clear Escalation Paths: When drift is detected, there must be a defined process for escalation. Who decides whether the model is recalibrate, retrained, or withdrawn? Without clarity, models may remain in use even as errors mount. Boards should demand documented response plans, with accountability assigned to named roles.

Vendor Oversight: For third-party models, drift management is partly outsourced. Contracts should include audit rights, performance reporting, and obligations to notify clients of significant retraining or degradation.

Black Box

Artificial intelligence is powerful not only for what it produces, but for what it hides. At the center of every AI system sits an algorithm: a set of rules or instructions that transform inputs into outputs.[30] In traditional software, those rules are written by humans and can be explained. A developer can point to the "if/then" statements or logic trees and show exactly how the program works.

In modern AI, particularly deep learning, those rules are not written line by line. They are learned by the system itself, through vast layers of statistical weights and probabilities. Inputs flow through millions or billions of parameters until an output emerges.[31] The result is a black box. We see the input and the output, but the reasoning in between is hidden, even from the engineers who built it.

This opacity is not just a technical quirk. It is a governance problem. If an organization cannot explain how an AI reached a decision, how can it trust it? How can it correct it? How can it defend it in court, or justify it to regulators?

The risks are not hypothetical. Recruitment algorithms have filtered out female applicants because they were trained on male-dominated data. Credit scoring systems have penalized applicants based on their postcode, effectively using geography as a proxy for race or income. Sentencing tools have recommended harsher penalties for minority groups. Each of these cases shared one feature. This is a hidden assumptions inside a model which produces visible harm outside it. And when challenged, the organizations could not explain how the decision was made.

If a system makes a discriminatory or harmful decision, who is responsible? The vendor that built it? The data scientists who deployed it?

Or the business that relied on it? Governance does not allow accountability to be outsourced. Directors remain responsible for outcomes, whether or not the logic is visible. That is why contracts with vendors must move beyond performance metrics to include obligations for explainability, audibility, and transparency.

Some AI models are inherently easier to interpret than others. Decision trees and regression models can be explained. Deep neural networks resist interpretation. Efforts in "explainable AI" are progressing, offering tools that highlight variables or build simplified surrogate models, but these are partial approximations. At best they provide glimpses, not a complete logic trail. Boards must distinguish between low-stakes automation, like drafting emails, and high-stakes decisions, like approving mortgages or diagnosing illness. In high-stakes contexts, explainability is not optional.[32]

Effective governance demands clarity about where AI operates and how its decisions are controlled. Boards must ensure there are fallback processes, especially if the outcome cannot be explained. Customers should be informed when AI influences their results and have clear channels to challenge or appeal those decisions. Regular audits of outputs for bias and accuracy build transparency, while contracts with vendors must protect the organization's right to understand how models perform.

Performance without explanation is fragile. A doctor who cannot explain a diagnosis erodes patient trust, even if correct. A bank that cannot explain a loan rejection erodes customer trust, even if compliant. Trust in institutions rests not only on outcomes but on accountability.

AI's power lies in its complexity, but so does its danger. Black box models deliver impressive outputs, but they obscure the reasoning that boards, regulators, and customers need to see. Without transparency, organizations cannot govern. Without governance, they cannot defend. And without defense, they cannot sustain trust. The lesson is simple: if you cannot explain an AI decision, you cannot own it.

Explainability Tools: Invest in explainable AI (XAI) methods such as SHAP or LIME that highlight which variables influenced a decision.[33] While imperfect, they give boards and regulators a basis for oversight.
Audit and Testing: Require regular third-party audits of model outputs. Independent testing can uncover bias or errors that internal teams miss. Boards should demand periodic reports and see trends over time.
Fallback Processes: or critical outcomes, insist on human-in-the-loop review. Customers denied a loan, job, or service must have an appeal pathway where a human can review and overturn an AI decision.
Contractual Safeguards: When relying on vendors, contracts must specify obligations for transparency, audibility, and cooperation with regulators. Boards should ensure that liability for opaque outcomes does not sit entirely with the enterprise.
Governance Alignment: Treat explainability as a regulatory and cultural expectation, not a technical extra. Ensure policies, risk frameworks, and training reinforce the principle that no important decision should be unexplainable.
Decision Documentation: Even when AI outputs appear accurate and fair, they may be challenged long after the decision is made. To defend those outcomes, organizations should maintain clear records of what data was entered, what the model produced, who reviewed the result, and why the final decision was taken. Documentation will not make a black box transparent, but it creates an evidential trail.

Data Poisoning

Artificial intelligence does not operate in isolation. It is only as strong as the data it is trained on, and that dependency creates one of its most dangerous vulnerabilities. Data poisoning is a form of quiet sabotage. Models are not compromised through code, but through corrupted inputs.[34]

Models are trained on vast datasets of text, images, audio, or records.[35] If those inputs are tainted, the intelligence that emerges will be tainted too. Poisoning occurs when malicious or flawed data is deliberately inserted into training. On the surface it looks ordinary, but it reshapes behavior in ways that are difficult to detect and impossible to reverse.

A poisoned model may reinforce bias, leak sensitive data, fail under certain conditions, or contain hidden triggers that can be exploited later.[36] What makes this so dangerous is its subtlety. No alarms go off. The system appears to work, but its decisions are warped and only reveal their flaws when it matters most.

Researchers have shown that inserting a small set of manipulated images into training data can cause stop signs to be misread as speed-limit signs. Language models have been primed with hidden instructions that remain dormant until triggered. In finance, subtle poisoning could let fraudulent transactions slip through. In healthcare, manipulated scans could cause missed diagnoses.

Attackers only need to insert a handful of poisoned records. Defenders must monitor billions. Big providers like Microsoft, Google, and AWS employ rigorous vetting, red-teaming, and access controls, but smaller vendors and in-house teams often rely on scraped or open-source data without safeguards. In those environments, poisoned inputs can pass unnoticed, embedding lasting vulnerabilities.

The issue is not linked to the mechanics but trust, liability, and oversight. An organization that cannot guarantee clean training data cannot guarantee safe outputs. As such, customer trust, investor confidence, and regulatory standing are challenged. Accountability does not vanish into the algorithm. If a poisoned model produces discriminatory or negligent outcomes, the liability rests with the organization that deployed it.

As many organizations choose to procure AI rather than build it, vendor dependency introduces new layers of risk. When suppliers rely on unverified or compromised datasets, that risk flows downstream to the business and its customers. Boards must ensure that management has strong oversight across the supply chain. Datasets need to be sourced, validated, and protected with the same rigour applied to internal systems. Processes for detecting data contamination should be in place, and vendor contracts must bind third parties to maintain data integrity.

AI is only as reliable as the integrity of its inputs. Poisoning exploits the opacity of AI, turning its greatest strength, learning from vast amounts of data, into a vulnerability. Every dataset is a potential attack surface, every vendor a potential weak link, and every poisoned model not just a technical flaw but a governance failure. Safeguards are not purely technical; they are contractual, cultural, and strategic. Clean data must be treated as a critical asset. Vendor agreements must emphasize transparency, not just speed. Oversight must elevate poisoning to the same level of concern as cybersecurity breaches, financial misstatements, or compliance violations.

The black box hides how models make decisions. Data poisoning corrupts what they learn in the first place.[37] Both remind us that AI is not magic. It is infrastructure, and an infrastructure that is compromised at the foundation will fail at scale.

Poisoning is not a research risk.It is already a tested exploit. [38] If data is the fuel of AI, boards must insist on guarantees that the fuel supply has not been tampered with. That requires oversight of how data is sourced, validated, and monitored, both in-house and through vendors. Lets look at several controls that can help mitigate this risk.

Data Provenance and Versioning: Track where training data comes from, how it is transformed, and when it changes. Tools such as ML-BOM or dataset version control can provide an audit trail. If poisoning is suspected, being able to roll back to a known-clean version is essential.

Vendor Accountability: Most organizations do not train models from scratch, they buy them. That makes vendor contracts critical. Agreements should specify how vendors source and validate data, what safeguards they use against poisoning, and whether independent audits are allowed. Without contractual guarantees, the risk of inherited poisoning is high.

Dataset Organization and Filtering: Training data should never be treated as a free commodity scraped off the web. Filters must be applied to screen for toxic, biased, or anomalous content. In smaller vendors or in-house builds, this step is often skipped for speed. Boards should ensure management resists that temptation.

Red Teaming and Stress Testing[39]: Models should be deliberately tested with adversarial inputs to see how they respond. Red-teaming exposes hidden triggers, back-doors, or weaknesses before attackers exploit them. Testing is not a one-off event but a continuous discipline.

Segregation of Critical Models[40]: Not all models are equal. A chatbot that drafts marketing copy carries less risk than a model that screens financial transactions. Boards should demand stricter controls on training data for models that influence critical or regulated decisions.

Supply Chain Exploit

Artificial intelligence is only as strong as the layers it is built on. Behind the polished interfaces of AI systems lie datasets, model weights, and architectures often stitched together from sources no one has fully verified. Enterprises spend heavily on firewalls, access controls, and endpoint security, yet the AI pipeline itself remains fragile and opaque.[41]

Unlike traditional supply chains, there is no global certification to guarantee the provenance of training data. No regulator audits model weights for tampering. Vendors face no binding requirement to disclose what data was used or how it was obtained. The result is an ecosystem built more on trust than assurance, more on speed than scrutiny.

This risk becomes sharper when AI development is outsourced.[42] Many businesses rely on third-party vendors to build models for analysis, customer interaction, or automation. But if those vendors are drawing on biased datasets, illegally scraped content, or malware-laced weights, then the risk is imported wholesale into the enterprise. When the development process is opaque, the model's decisions cannot be trusted.

Security researchers compare data poisoning in the AI pipeline to inserting compromised firmware into a motherboard. The attack is quiet, subtle, and persistent. A poisoned dataset can skew outputs, reinforce stereotypes, or embed hidden back-doors that trigger under specific conditions. A tampered weight file can alter how a model classifies inputs. None of this requires brute force. It enters unnoticed, and the damage only becomes clear later when outputs cause compliance breaches, reputational harm, or financial loss.

In August 2025, attackers compromised Salesloft's Drift, an AI-driven chatbot integrated with Salesforce.[43] Using stolen OAuth tokens, they gained access to hundreds of Salesforce environments, extracting sensitive data, customer records, and even credentials like AWS keys. The breach cascaded to firms as prominent as Zscaler, Cloudflare, and Palo Alto Networks. The flaw was not in Salesforce itself but in the AI integration. This is a vivid reminder that third-party tools are part of the supply chain, whether boards acknowledge them or not.

AI supply chain exploits expose organizations to operational, regulatory, and reputational risk.[44] Compromised datasets or integrations can corrupt decision-making across finance, HR, and customer service. Regulators will impose penalties if training data is illegally sourced or if bias seeps into outputs under GDPR, the EU AI Act, or the *Privacy Act*. Yet the AI supply chain is far less robust than many assume. Enterprises may spend millions on access controls, detection, and firewalls, but the foundations of AI, including datasets, weights, pre-trained models, and fine-tuning adapters, are often stitched together from sources few have fully vetted.

Unlike traditional supply chains, there is no global standard for model provenance. Vendors are not required to certify where training data comes from or disclose whether open-source components or adapters have been altered. Too often, AI is built on speed and trust rather than scrutiny and assurance.

Every dataset, plugin, and adapter forms part of the AI supply chain, whether recognized or not. Customers do not distinguish vendor failure from brand failure. Yet few organizations can trace the origin of their datasets or model weights, or verify that independent audits were conducted. Boards must insist on transparency. Procurement should require disclosure of data sources, audit results, and contractual guarantees of integrity, applying the same scrutiny used for hardware, cloud, or other critical dependencies.

Just as regulators demand traceability from farm to fork, enterprises must demand it from dataset to deployment. AI systems are only as secure as their weakest supplier. A poisoned dataset, tampered adapter, or vulnerable integration can quietly compromise operations. These may not look like conventional breaches, but their outcomes, such as fines, reputational damage, and systemic failure, are just as serious. Let's look at a few controls which can help reduce this risk.

Provenance and Disclosure: Boards should demand transparency about where training data, pre-trained models, and fine-tuning adapters originate. Procurement processes must require disclosure of dataset sources, model provenance, and licensing conditions. Just as food supply chains track origin and handling, AI models should be traceable from dataset to deployment.

Audit and Independent Assurance: Enterprises should not take vendor claims at face value. Independent audits, red-team testing, and adversarial evaluations are critical. Vendors must be able to demonstrate that their models and adapters have been stress-tested against poisoning, tampering, and bias. Where possible, insist on third-party assurance reports.

Bill of Materials (SBOM / MBOM): Boards should expect an inventory of components that make up a model. A Software Bill of Materials (SBOM) or Model Bill of Materials (MBOM) allows organizations to see which datasets, libraries, or adapters are in use. This makes it easier to identify vulnerabilities when suppliers issue patches or when malicious components are discovered.

Vendor Contracts and Accountability: Contracts must include obligations for provenance, explainability, and data integrity. Vendors should provide guarantees of patching support, incident notification, and liability if poisoned or tampered models are supplied. Without contractual teeth, assurances are little more than marketing promises.

Monitoring and Testing: Supply chain exploits thrive in opacity. Continuous monitoring is required to detect unusual behaviors in vendor models. Scenario testing, anomaly detection, and adversarial probes should be used to validate whether a model continues to behave as expected. This should be built into MLOps pipelines, not left as an afterthought.

Patch and Update Discipline: Outdated models and libraries are a known risk. Enterprises must ensure that vendors maintain active patching cycles and that integrations are updated regularly. Boards should ask for evidence of patching policies and confirmation that deprecated or unsupported models are not embedded in business-critical systems.

Access and Integration Controls: Every integration,whether a chatbot, API, or SaaS plugin, is part of the AI supply chain. Strong access controls (OAuth hygiene, MFA, monitoring for token theft) and network segmentation can limit the blast radius if one integration is compromised. The Drift breach shows how a single weak link can cascade across hundreds of environments.

The Missing Value (NaN)

The collection of data often faces quality issues. Records arrive incomplete, forms are left half-filled, sensors drop readings, and human error introduces gaps. In practice, these gaps are commonly marked as NaN (Not a Number).[45] At first glance, they seem minor. Yet when they accumulate at scale, they can quietly destabilize entire models.

Missing values remain a persistent problem. Think about the last time you filled out an online form. Did you complete every field? And if you did, did you provide identical information each time? Data quality continues to be a foundational concern, particularly in privacy. Under the Australian *Privacy Act* 1988, Australian Privacy Principle 10 explicitly requires that personal information be accurate, complete, and up to date.

When working with datasets that include NaN values, often treated as empty cells, choices must be made. Should missing values be filled, dropped, or ignored? Each decision reshapes the dataset in subtle but significant ways. Filling gaps with averages may smooth the data but erase important distinctions.

Dropping rows may remove noise but also reduce diversity, leading to skewed results. Ignoring NaNs can cause silent malfunctions or introduce hidden biases. These choices are frequently buried deep within preprocessing pipelines, set as defaults, or left undocumented. Yet their downstream effects reach into business strategy, compliance, and risk exposure. What seems like a small technical decision can, in the end, influence entire markets.

In 2021, Zillow shut down its Zillow Offers business after losing more than $500 million in a single quarter.[46] At the center was the company's

Zestimate algorithm, which overestimated home values and led to systematic overpayment. Housing markets are messy and many property details were missing. Data such as the quality of renovations, the condition of interiors, the dynamics of particular streets, all have an impact. To compensate, the model filled gaps with averages or dropped incomplete records. On dashboards, the model looked confident. In reality, the gaps compounded, errors widened, and the business collapsed. Missing values are not just statistical noise. They can scale into existential risk.

NaNs also carry a bias problem. Missing values are rarely random. They cluster around particular groups or conditions.[47] In healthcare, disadvantaged patients may have incomplete medical records, so dropping those rows erases vulnerable populations from the dataset.

In finance, applicants without long credit histories generate more NaNs, and filling with averages unfairly penalizes younger or minority borrowers. In employment, non-traditional resumes may lack standardized data fields, causing models to overlook otherwise qualified candidates. In each case, mishandling NaNs does not just reduce accuracy, it produces outcomes that disproportionately harm those already at the margins.

The operational and compliance risks are equally sharp. Models that treat NaNs inconsistently produce unstable outputs. Smoothing over gaps can inflate accuracy metrics, giving a false sense of confidence. Systems may run without error messages, but their decisions are distorted. In regulated industries, this translates into liability. A credit model that mishandles missing values can breach fair lending rules. A healthcare AI that drops incomplete records can expose providers to claims of discriminatory care. Regulators will not accept "the dataset had gaps" as an excuse.

NaNs illustrate that AI risk does not reside only in algorithms but also in the data practices that support them. When the governance of preprocessing is weak, the entire model becomes fragile. Effective oversight requires visibility into how missing values are managed and whether there is a consistent, documented policy for handling them. Decisions about imputation, exclusion, or substitution must be reviewed and auditable, not left to individual discretion.

Unlike many emerging risks, NaNs are not mysterious. Data scientists know how to handle them. The gap is not technical knowledge, it is governance discipline. Too often, missing values are dismissed as a technical detail beneath board concern. Zillow's collapse shows the danger of that thinking. Boards must insist on enterprise standards. Preprocessing choices should be documented and auditable. In high-stakes contexts such as finance, healthcare, and HR, these decisions should be escalated to governance review. Regular validation must show not only how NaN handling affects accuracy, but also fairness and compliance.

Missing values are not trivial technical details. They shape how models see the world and how decisions are made at scale. Small decisions in preprocessing can lead to serious outcomes, including bias, compliance failures, or financial loss.

Treating NaNs as a governance issue means demanding visibility, setting clear standards, and ensuring accountability in how data gaps are addressed. With that in mind, let explore a few controls that can reduce this risk.

Standardization of Data Handling: Define enterprise rules for when to impute, drop, or escalate missing values. Inconsistent handling across teams creates fragmented outcomes and hidden risks.

Transparency, Oversight and Risk Assessments: Every reprocessing decision should be documented, reviewed, and justified through a simple risk assessment. Dual sign-off for critical datasets ensures accountability. Boards should expect clear reporting on how missing values are managed and that related risks have been assessed and mitigated.

Fairness Testing: NaNs often cluster around specific groups, such as disadvantaged patients or younger applicants. Regular fairness testing should confirm that data handling does not harm vulnerable groups.

Validation Against Outcomes: Imputation techniques must be tested against real-world results. For instance, does averaging distort risk scores? Does dropping incomplete resumes bias recruitment?

Escalation of Critical Decisions: In regulated or high-impact domains, preprocessing rules must be reviewed by governance teams when they affect finance, healthcare, HR, or legal contexts.

Automation with Guardrails: Automated tools may process routine NaNs, but they should include alerts and thresholds to flag unusual gaps before outputs are trusted.

Vendor Oversight: Contracts with external providers must require disclosure of how missing values are handled and grant audit rights to verify compliance.

Training and Awareness: Leaders and staff must understand that missing data is not trivial. Training should highlight real cases, such as Zillow, where poor NaN handling caused financial loss.

Over-Reliance

As AI becomes embedded in operations, people can grow over-reliant on it. Confidence in personal judgment fades, replaced by misplaced trust in machine outputs. When AI delivers inaccurate or misleading information, employees may hesitate to act independently, creating a cycle of dependency that weakens decision quality and exposes the organization to disruption.

Real-world cases show the risk. Lawyers filed court documents drafted by ChatGPT that cited cases which did not exist. A police system wrongly identified an individual, leading to a wrongful arrest. A teenager tragically took their own life after following AI-generated advice. In each case, the issue was not malicious AI but human deference to flawed outputs.

The impacts are significant. Lost capability slows productivity. Unchecked errors cause reputational and legal harm. Correcting flawed data is costly. Dependence limits creativity and agility. And deep system integration can lead to costly vendor lock-in.

AI should augment judgment, not replace it. Controls must help staff retain critical skills, include verification steps, and provide fallbacks if systems fail. Training should emphasize both the strengths and limits of AI. Oversight metrics should track not just system accuracy but also the level of human review applied.

Over-reliance is not inevitable. When people stay involved, question outputs, and treat AI as a partner rather than a replacement, the result is more informed, reliable, and resilient decisions. Let's look at a few controls to help reduce this risk.

Maintain Human Oversight: AI outputs should never be the final word. Workflows must include review points where humans validate recommendations, check for errors, and exercise their own reasoning. This ensures the system supports staff rather than replacing them.

Training and Awareness: Training programs should include manual processes to reduce the risk of 'skill atrophy'. Maintaining these capabilities builds resilience if AI tools fail or need to be withdrawn.

Establish Fallback Procedures: Processes that rely on AI should have a documented backup plan. If an AI tools are unavailable, staff must know how to continue operations.

Exit Strategy: AI systems should have an exit strategy. Even if contracts allow for termination, dependency on AI could make exit difficult. If processes, data, and staff practices are fully embedded into a single vendor's tools, the business may find itself locked in to a service they technically cannot leave. Boards should insist on transition planning, portability of data, and alternative pathways before systems become too entrenched.

Human Factor Risk

A major IT outage occurred when an employee removed what appeared to be an unused network link. The link was part of a wider dependency chain, and within minutes, all users and devices were locked out of the system. Acting with good intent, the individual followed instructions generated by ChatGPT and made the change without expert validation. The result was complete network inaccessibility and business disruption.[52]

This incident shows how easily systems can fail when human decision-making is guided by unverified AI outputs. The issue was not malicious but it illustrates a breakdown in governance, oversight, and procedural control. Human Factor Risk arises when well-meaning staff rely on tools or instructions they do not fully understand, without the guardrails of supervision or peer review.

Human Factor Risk reminds boards that even in an AI-assisted environment, responsibility cannot be handed to technology. Training, change management, and expert validation remain essential safeguards against preventable harm. While it overlaps with Over-Reliance Risk, the distinction lies in scope and impact. Over-Reliance occurs when humans refer to AI outputs without validation, leading to flawed decisions or reputational harm. Human Factor occurs when human or AI-guided action directly disrupts systems, affecting integrity and continuity.

Both risks stem from misplaced confidence and weak governance. Over-Reliance damages trust, while Human Factor damages function. Managing both requires training, validation, and layered oversight, but with different points of focus. Over-Reliance requires scrutiny of outputs; Human Factor requires control over inputs and actions. Let's look at a few mitigants that can reduce this risk.

Access Management: Ensure there is tiered access management in place, where users are granted the minimum level of system access necessary for their role. This includes defining role-based permissions, enforcing strong authentication mechanisms, and regularly reviewing and updating access rights to prevent unauthorized changes or mistakes.

Change Management: Implement strict change management processes that require proper planning, approval, testing, and documentation for all system configuration changes. This ensures that all changes are performed by qualified personnel and reviewed to minimize the risk of operational disruptions or security incidents.

Human Resources and Recruiting: Recruitment should prioritize clarity and verification. Roles involving system or privileged access must be advertised with clear technical expectations, required certifications, and decision-making scope. Before granting privileged access, evidence of current technical certification or verified internal proficiency should be obtained and documented.

System Safeguards and Guardrails: Systems should include built-in protections that prevent unvalidated or high-risk actions, such as automatic dependency checks, access alerts, and rollback capabilities. These technical guardrails act as a safety net when human error or process gaps occur, helping to maintain stability and recover quickly from mistakes.

Configuration and Monitoring: Establish and enforce secure baseline configurations using Group Policy Objects (GPOs) and regular configuration reviews. Continuous monitoring should detect unvalidated or misconfigured changes early, ensuring that configuration drift or incorrect system modifications do not cascade into widespread failure.[53]

Recruiting Risk

AI adoption can influence talent acquisition. Developers, data scientists, and other skilled professionals value the tools and practices they work with. If a company commits to a model or vendor seen as unsafe, outdated, or misaligned with industry standards, candidates may decline an offer. Reputation matters in attracting high-demand talent.

Developers and technical staff may assess potential employers by the tools and models they use. Outdated systems signal inefficiency and immaturity, damaging credibility and deterring the expertise needed to advance AI initiatives. In competitive labour markets, this can leave organizations understaffed or over-reliant on existing teams.

In an AI-driven environment, an unattractive or ethically questionable AI stack will cause top talent to walk away regardless of salary. Governance must therefore extend to talent acquisition. The following controls could help mitigate this risk.

Updated Tools: Recruitment is not always the answer to business challenges. Leaders should first ask why they are hiring. Often, the issue may not lie in the headcount but in outdated tools or inefficient systems. Hiring more people will not fix a broken process. Upgrading technology or streamlining workflows is often the faster and more cost-effective path.

Technology Roadmap and Assessment:A clear roadmap helps prevent recruitment and reputational risks caused by outdated tools. It should show which models and systems are in use, when they will be updated, and how new technologies are evaluated. The roadmap must link to the organization's AI risk assessment so adoption decisions balance innovation with responsibility. Demonstrating this discipline reassures recruits, regulators, and partners that the business is advancing technology in a controlled and transparent way.

AI Risk and Board Accountability

AI governance always leads back to something human: accountability. Boards cannot pass that responsibility to data scientists, vendors, or regulators. Oversight sits with directors, and in the age of AI, that duty is expanding.

Real oversight means more than polished dashboards or vendor briefings. Boards need to understand how AI models are monitored in real time, what thresholds trigger retraining or withdrawal, and who can pause or retire a model that drifts off course.

Transparency is critical when performance declines. Problems cannot be buried within technical teams. In high-stakes environments, ignorance is not a defense. Regulators, courts, and customers will not accept "we did not know" when harm occurs.

AI systems do not operate in isolation. They need regular monitoring, retraining, and feedback loops to detect drift. Human oversight is essential when outcomes affect people's lives, money, health, or liberty. A model that drifts is not just a technical failure; it is a governance failure. The board's responsibility does not end when a project is approved. It continues as the system learns, adapts, and interacts with the world.

That responsibility extends beyond the enterprise. Very few organizations build AI in-house. Most rely on vendors, pre-trained models, or cloud platforms. Each forms part of the AI supply chain, and each introduce risks that trace directly back to the board.

Contracts must reflect reponsibility. They should define accountability for performance, require vendors to disclose how models are monitored and retrained, and set clear terms for liability if degraded or poisoned models cause harm. Vendors must also be required to notify customers when AI is embedded in their services or when its function materially changes.

Without this transparency, boards cannot measure exposure or meet regulatory obligations. AI does not remove responsibility. It expands it.

AI should be treated as infrastructure. Like roads, energy grids, or financial systems, it underpins how an organization functions. And like all infrastructure, it carries risk if it is poorly maintained, poorly monitored, or poorly governed. Boards must see AI not as a passing trend but as a standing governance duty. That means asking direct questions, demanding clarity, and expecting proof that controls are working.

Finally, boards must avoid the temptation to chase every new AI tool. Pressure to adopt systems too quickly, driven by fear of missing out or promises of efficiency, can lead to poor decisions. Without due diligence, enthusiasm can turn into exposure, inviting regulatory scrutiny, financial loss, or reputational harm. The power to approve must be exercised with care, ensuring that innovation never outweighs accountability.

Reference

1. Alan M Turing, 'Computing Machinery and Intelligence' (1950) 59(236) Mind 433.
2. Lora Shinn, 'Voice Cloning AI Scams Are on the Rise' (BECU Blog, 29 September 2025) https://www.becu.org/blog/voice-cloning-ai-scams-are-on-the-rise.
3. Matt Martino and Devi Mallal, 'We Cloned Senator Jacqui Lambie's Voice with AI to Show What a Deepfake Election Could Look Like' ABC News (27 February 2025) https://www.abc.net.au/news/2025-02-28/jacqui-lambie-ai-generated-voice-election-and-deepfakes/104986434.
4. Laura Goodwin, 'Romance Scammer Duped £17k from Me with Deepfakes' BBC News (2024) https://www.bbc.com/news/articles/cdr0g1em52go.
5. US Department of Homeland Security, Increasing Threat of Deepfake Identities (2021) 5–7 https://www.dhs.gov/sites/default/files/publications/increasing_threats_of_deepfake_identities_0.pdf.
6. Sonam Singh and Amol Dhumane, 'Unmasking Digital Deceptions: An Integrative Review of Deepfake Detection, Multimedia Forensics, and Cybersecurity Challenges' (2025) MethodsX 15.
7. Emily Bell, 'A Fake News Frenzy: Why ChatGPT Could Be Disastrous for Truth in Journalism' The Guardian (3 March 2023) https://www.theguardian.com/commentisfree/2023/mar/03/fake-news-chatgpt-truth-journalism-disinformation.
8. Sarah Tait, 'Latest Trends of Generative AI on the Side of Criminals' (KPMG UK, 2025) https://kpmg.com/uk/en/insights/ai/cdd-and-generative-ai-risks-of-kyc-fraud.html.
9. Ellen Breugem, 'Fraud Attempts with Deepfakes Have Increased by 2137% Over the Last Three Years' (Signicat, 2025) https://www.signicat.com/press-releases/fraud-attempts-with-deepfakes-have-increased-by-2137-over-the-last-three-year.
10. US Department of Homeland Security, Increasing Threat of Deepfake Identities (2021) 5–7.
11. Dan Milmo, 'Company Worker in Hong Kong Pays Out £20m in Deepfake Video Call Scam' The Guardian (5 February 2024) https://www.theguardian.com/world/2024/feb/05/hong-kong-company-deepfake-video-conference-call-scam.
12. Jess Daniel, 'A Voice Deepfake Was Used to Scam a CEO Out of $243,000' Forbes (3 September 2019) https://www.forbes.com/sites/jessedamiani/2019/09/03/a-voice-deepfake-was-used-to-scam-a-ceo-out-of-243000/.
13. Daniel Jurafsky and James H Martin, Speech and Language Processing (3rd ed, draft, Prentice Hall, 2023).

14. Govind Mittal et al, 'Pitch: AI-Assisted Tagging of Deepfake Audio Calls Using Challenge–Response' (ASIA CCS '25, Hanoi, 25–29 August 2025) ACM https://doi.org/10.1145/3708821.3733861.
15. Privacy Act 1988 (Cth); Regulation (EU) 2016/679 (GDPR) [2016] OJ L119/1.
16. James Vincent, 'An Online Propaganda Campaign Used AI-Generated Headshots to Create Fake Journalists' The Verge (2020) https://www.theverge.com/2020/7/7/21315861/ai-generated-headshots-profile-pictures-fake-journalists-daily-beast-investigation.
17. Kai-Cheng Yang, Danishjeet Singh and Filippo Menczer, 'Characteristics and Prevalence of Fake Social Media Profiles with AI-Generated Faces' (2024) arXiv preprint https://arxiv.org/abs/2401.02627.
18. Gae Regan, 'How Deepfakes Exploit KYC Verification Systems' (Reality Defender, 2025) https://www.realitydefender.com/insights/how-deepfakes-exploit-kyc-verification-systems.
19. Microsoft, Data, Privacy, and Security for Microsoft 365 Copilot (Microsoft Learn) https://learn.microsoft.com/en-us/copilot/microsoft-365/microsoft-365-copilot-privacy.
20. Aditya Patel, 'AI Gone Wild: Why Shadow AI Is Your IT Team's Worst Nightmare' (Cloud Security Alliance Blog, 4 March 2025) https://cloudsecurityalliance.org/blog/2025/03/04/ai-gone-wild-why-shadow-ai-is-your-it-team-s-worst-nightmare.
21. Aaron Mok, 'Google employees who tested its ChatGPT-rival Bard reportedly called it "cringeworthy" and a "pathological liar," and said the AI's responses could "result in serious injury or death"' (19 April 2023) Business Insider https://www.businessinsider.com/google-workers-called-chatgpt-ai-rival-bard-cringworthy-liar-report-2023-4
22. Michele Salvagno, Fabio Silvio Taccone and Alberto Giovanni Gerli, 'Artificial Intelligence Hallucinations' (2023) 27(1) Critical Care 180.
23. Daniel Jurafsky and James H Martin, Speech and Language Processing (3rd ed, draft, Prentice Hall, 2023).
24. Damien Carrick and Sophie Kesteven, 'This US Lawyer Used ChatGPT to Research a Legal Brief with Embarrassing Results. We Could All Learn from His Error' ABC News (24 June 2023) https://www.abc.net.au/news/2023-06-24/us-lawyer-uses-chatgpt-to-research-case-with-embarrassing-result/102490068.
25. "Hallucinations in AI-generated medical summaries remain a grave concern" (2024) https://www.clinicaltrialsarena.com/news/hallucinations-in-ai-generated-medical-summaries-remain-a-grave-concern/.clinicaltrialsarena.com/news/hallucinations-in-ai-generated-medical-summaries-remain-a-grave-concern/.

26. Negar Maleki, Balaji Padmanabhan and Kaushik Dutta, 'AI Hallucinations: A Misnomer Worth Clarifying' (2024) IEEE Conference on Artificial Intelligence (CAI).
27. Surya Patchipala, 'Tackling Data and Model Drift in AI: Strategies for Maintaining Accuracy During ML Model Inference' (2023) 10(2) International Journal of Science and Research Archive 1198.
28. Ylenia Rotalinti et al, 'Detecting Drift in Healthcare AI Models Based on Data Availability' (2022) Joint European Conference on Machine Learning and Knowledge Discovery in Databases (Springer).
29. Ibid.
30. Adriana Sayuri Iwashita and Joao Paulo Papa, 'An Overview on Concept Drift Learning' (2018) IEEE Access 7, 1532.
31. Valerie du Preez et al, 'From Bias to Black Boxes: Understanding and Managing the Risks of AI — An Actuarial Perspective' (2024) British Actuarial Journal 29, e6.
32. Vladislav V Fomin and Paulius Astromskis, 'The Black Box Problem' in Future Law, Ethics, and Smart Technologies (2023) 112–125.
33. Stuart J Russell and Peter Norvig, Artificial Intelligence: A Modern Approach (4th ed, Pearson, 2020).
34. Sajid Ali, et. al, Explainable Artificial Intelligence (XAI): What we know and what is left to attain Trustworthy Artificial Intelligence (2023) 99 Information Fusion 101805.
35. Halima I Kure et al, 'Detecting and Preventing Data Poisoning Attacks on AI Models' (PhotonIcs & Electromagnetics Research Symposium, Abu Dhabi, 4–8 May 2025).
36. Daniel Jurafsky and James H Martin, Speech and Language Processing (3rd ed, draft, Prentice Hall, 2023).
37. Halima I Kure et al, 'Detecting and Preventing Data Poisoning Attacks on AI Models' (2025).
38. OWASP, Data and Model Poisoning (2025) https://genai.owasp.org/llmrisk/llm042025-data-and-model-poisoning/.
39. OWASP, Data and Model Poisoning (2025) https://genai.owasp.org/llmrisk/llm042025-data-and-model-poisoning/.
40. OWASP, Supply Chain (2025) https://genai.owasp.org/llmrisk/llm032025-supply-chain/.
41. OWASP, OWASP Top 10 for Large Language Model Applications (2025) https://owasp.org/www-project-top-10-for-large-language-model-applications/

42. Ibid
43. Ibid.
44. Sead Fadilpašić, 'Zscaler Says It Suffered Data Breach Following Salesloft Drift Compromise' TechRadar (2 September 2025) https://www.techradar.com/pro/security/zscaler-says-it-suffered-data-breach-following-salesloft-drift-compromise.
45. Utkarsh Mittal, 'AI-Based Evaluation System for Mitigating Supply Chain Risk' Authorea Preprints (2023).
46. Ian Goodfellow, Yoshua Bengio and Aaron Courville, Deep Learning (MIT Press, 2016).
47. Ryan Dezember, 'Zillow's Flawed Algorithm: How an AI Model Misread the Housing Market' Wall Street Journal (12 November 2021).
48. Yiliang Zhang and Qi Long, 'Fairness in Missing Data Imputation' (2021) arXiv preprint.
49. Ian Goodfellow, Yoshua Bengio and Aaron Courville, Deep Learning (MIT Press, 2016).
50. Microsoft, 'Fostering Appropriate Reliance on GenAI — Overreliance on AI: Risk Identification and Mitigation Framework' (Microsoft Learn, 4 March 2025) https://learn.microsoft.com/en-us/ai/playbook/technology-guidance/overreliance-on-ai/overreliance-on-ai.
51. Artur Klingbeil, Cassandra Grützner and Philipp Schreck, 'Trust and Reliance on AI — An Experimental Study on the Extent and Costs of Overreliance on AI' (2024) 160 Computers in Human Behavior 108352 https://doi.org/10.1016/j.chb.2024.108352.
52. Nadine Yousif, 'Parents of Teenager Who Took His Own Life Sue OpenAI' BBC News (2025) https://www.bbc.com/news/articles/cgerwp7rdlvo.
53. F3ndt, 'Emergency Help – Entire Domain Inaccessible' (Reddit, 7 hours ago) https://www.reddit.com/r/sysadmin/comments/10jbifu/emergency_help_entire_domain_inacessible/
54. Microsoft Corporation. (Year). Title of document: Remove orphaned domains from Active Directory [Web page]. Microsoft. Retrieved Month Day, Year, from https://learn.microsoft.com/en-us/troubleshoot/windows-server/windows-security/remove-orphaned-domain
55. Paul Daugherty and H James Wilson, Radically Human: How New Technology Is Transforming Business and Shaping Our Future (Harvard Business Review Press, 2022).

"In security, AI is both the lock and the crowbar."

CHAPTER 9

AI Security - The Big Four

Security has always been the 'silent' architecture beneath every system of trust. It is the discipline that transforms technological possibility into operational reality. Without it, innovation collapses under the weight of its own complexity. Every digital interaction, from a bank transfer to a medical record update, depends on the assurance that data is authentic, private, and available when needed. Security is therefore not a feature to be added later but the condition that makes all other functions possible.[1]

The pursuit of security has mirrored the evolution of technology itself. Early computing focused on protecting mainframes and physical access, while the rise of distributed systems introduced new exposure points that demanded layered defense.[2] As networks extended beyond the corporate perimeter, the focus shifted from protecting machines to protecting information.[3] Yet the goal has never changed. Security's goal is to preserve the integrity of data and the continuity of operations. What has changed is the scale and speed of what requires defending.[4]

Everything is connected in the digital world. A single compromised endpoint can trigger a cascade that affects supply chains, financial markets, and national infrastructure.[5] Attackers exploit this connectivity through automation, social engineering, and artificial intelligence itself, often moving faster than human defenders can respond.[6] This environment has transformed cybersecurity from a technical safeguard into a business-critical capability that defines resilience, reputation, and competitiveness.

Security is also a measure of governance. It reflects how well an organization identifies assets, assesses risk, and prepares for disruption.[7] Strong governance does not remove threats, but it reduces uncertainty and creates accountability.[8] Boards that treat security as a cost rather than as a strategic investment often discover that it is the foundation of business continuity. In modern organizations, every control, policy, and safeguard represents capital protection as much as compliance.

The arrival of artificial intelligence adds a new dimension to this responsibility. AI systems learn and act at speeds beyond traditional oversight, drawing power from massive datasets and distributed infrastructure.[9] With this capability comes dependency. When the systems that interpret and act on our behalf are compromised, the impact is spread across decision-making, service delivery, and public trust. Security is therefore not only about defending data but about protecting the integrity of reasoning itself.

What makes AI security challenging is its subtlety. Attacks can occur at any stage of the model lifecycle from during data collection, training, or deployment, to interaction. Manipulated inputs can distort predictions while appearing legitimate.[10] Adversarial prompts can extract confidential information or introduce bias into results.[11] These are not just technical failures but governance breakdowns, demonstrating why oversight of AI must combine cyber, legal, and ethical disciplines.

Technology decisions are no longer matters delegated to IT teams. They are strategic questions about organizational safety. The same principles that protect networks must now extend to algorithms, datasets, and learning environments.[13] A failure in AI governance can have the same impact as a data breach or infrastructure outage. Security becomes the common thread linking ethics, compliance, and resilience.

Ultimately, security is the guardian of trust. It assures customers that their data is safe, employees that their contributions are valued, and investors that their assets are protected. In a world defined by automation and data exchange, trust is not earned through statements but through demonstrable control. Security provides that proof. It remains the silent measure of credibility in every transaction, partnership, and innovation.

As organizations begin to embed artificial intelligence into their operations, the meaning of security expands. The same principles that protect traditional systems now apply to models, data pipelines, and automated decision-making. Trust in AI is built in exactly the same way as trust in any digital system, through evidence of control, resilience, and accountability. What changes is not the purpose of security but the scale and complexity of what must be secured.

Securing AI

Artificial intelligence is only as strong as the environment in which it operates. Just as financial systems depend on audit and aviation depends on air traffic control, AI depends on a secure and resilient foundation. In boardrooms the conversation about AI security often drifts toward abstract concepts such as "trustworthy AI" or "responsible innovation." These ideas are important, but they are incomplete. What directors need is a clear view of the operational pillars that keep AI secure in practice.

I will discuss, what I have termed as, the Big Four: infrastructure readiness, network security, software security, and cloud security. Together they form the foundation that determines whether AI deployments are safe, sustainable, and resilient. Each pillar has its own risks, but they are connected. A weakness in one can quickly undermine the others. A resilient network means little if the underlying infrastructure is fragile. Strong software controls achieve little if the cloud environment is misconfigured. Cloud readiness cannot make up for poor organizational policies or unprepared data centers.

Boards and organisation should approach AI security as a systemic challenge. This requires moving beyond isolated fixes or vendor assurances and focusing on how these four domains work together. Infrastructure readiness ensures that the foundation of AI, including data centers, hardware accelerators, and energy systems, can support critical workloads without collapse. Network security governs the flow of data and ensures that AI is not exposed to interception, manipulation, or denial-of-service attacks. Software security protects the integrity of applications, APIs, and development practices that feed into AI systems. Cloud security ensures that the environment in which AI operates is properly configured, continuously monitored, and resilient against threats.

The Big Four provide a foundation for asking more direct questions. They help decision-makers see where vulnerabilities lie and where accountability sits. If AI is to become part of core business operations, directors must ensure that these four areas are not treated as afterthoughts but as strategic foundations.

Infrastructure Readiness

Artificial intelligence is not a browser with search features. It is not a simple add-on to existing systems, nor is it another application that staff can download and use without consequence. AI is built on and powered by infrastructure. It relies on data centers, networks, hardware accelerators, and organizational systems that determine whether it is secure, resilient, and reliable. This readiness is not about environmental footprint or ESG performance, though these are important in their own right. It is about the operational backbone that allows AI to function safely inside an enterprise.

Vendors often present "AI readiness" in broad, attractive terms. Microsoft, for example, has spoken about Copilot readiness in language that highlights governance, identity, and enablement.[12] These are important, but the details of which technical features must be configured, which controls must be enforced, and which risks must be mitigated are far less visible.

The term readiness is often misunderstood and can be misleading, as it implies the organization is safe to proceed when it may simply mean that preparations are in place. In reality, what many vendors describe is aspirational, not operational. The gap between high-level readiness and real security maturity is wide, and this gap is where organizations expose themselves most to risk. True readiness spans several layers, including physical, organizational, governance, and technical. Each layer requires its own safeguards.

Levels of readiness vary across organizations. Some organizations have invested in data classification but have no policies to ensure that classification is applied consistently.[13] Others have developed labeling frameworks but have not yet introduced data loss prevention. A few have implemented both classification and DLP(Data Loss Prevention), yet still have no policies that apply specifically to AI inputs and outputs. Many organizations may be in an even more fragmented position, with one business function relatively secure while another is entirely unprotected. As such, AI is being layered onto this uneven terrain, which magnifies weaknesses and creates pathways for misuse or error.

Readiness is not a single milestone but a condition that spans five layers: physical, organizational, governance, technical, and cloud. Each must be managed with the same discipline and scrutiny as any other critical business system.

The physical layer is the foundation. It includes data centers, GPUs, TPUs, and power systems that enable model training and operation. These assets are attractive targets for sabotage, ransomware, and supply chain disruption. Organizations that rely on AI for essential functions cannot assume existing data center resilience is sufficient. Boards should confirm that the resilience of compute environments matches the importance of the business functions they support. They should also ensure that continuity plans and tested recovery processes are in place so the organization can continue to operate if this layer fails.

Hardware supply chains present another vulnerability. The global demand for GPUs has shown how fragile AI's infrastructure can be. Export restrictions, natural disasters, or targeted interference can interrupt access to compute capacity overnight. Organizations that depend on AI must plan for these scenarios and know how operations will continue if those resources become unavailable.

The organizational layer is equally important. Policies, procedures, and defined roles are as essential as hardware. Without clear guidance on AI use, staff will experiment with tools such as Copilot or ChatGPT. This behavior is rarely malicious; it reflects curiosity and a desire to work more efficiently. Without boundaries, however, employees may upload sensitive information to external systems, creating compliance and privacy risks. Even where policies exist, inconsistent enforcement can leave blind spots that remain unnoticed until an incident occurs. Boards should ensure policies are communicated, understood, and consistently applied across all levels of the organization.

The governance layer links policy to action. Data classification, lifecycle management, and access controls connect what is written in policy to how systems are managed in practice. Too often, these mechanisms are uneven. Classification may exist without data loss prevention to enforce it, or protection measures may be limited to one part of the business. These gaps allow AI systems to access or expose data unintentionally. Governance is only effective when oversight and enforcement operate together. Boards should seek assurance that frameworks are not only documented but actively integrated into day-to-day operations.

The technical layer determines whether AI systems operate securely. Identity protection, conditional access, tenant isolation, logging, and data sovereignty are essential controls. Yet vendor material often describes a vision of security rather than specific actions. Many "readiness" guides focus on adoption rather than configuration. Boards should avoid accepting readiness claims at face value. They should request evidence that configurations have been tested and that technical safeguards match the organization's risk profile.

Copilot provides a clear example. When data classification and labelling are not applied, and data loss prevention is absent, the system can surface sensitive content from across the tenant, including restricted material. Although the data remains technically within the tenant, users may still see information they should not access. This can result in regulatory breaches and reputational harm.

The cloud layer underpins all others. It determines how AI systems scale, integrate, and exchange data. Misconfigured APIs, insecure storage, or weak access controls can turn a cloud platform into a single point of exposure. Shared responsibility models can add confusion, leaving gaps in accountability between vendors and internal teams. Boards should insist on clear ownership of each risk and verify that security settings in cloud environments are reviewed regularly.

Many organizations are not yet ready for enterprise-scale AI. Some have policies but little enforcement. Others apply data labelling inconsistently. Only a few have reached higher maturity, where classification, data loss prevention, and governance are fully aligned. The most advanced organizations monitor AI systems continuously, use adaptive controls, and embed AI-specific risk frameworks.

Inadequate preparation has real consequences. Deploying AI in immature environments exposes sensitive data and increases operational risk. Boards must know where their organization sits on the maturity spectrum. Investment in readiness is not optional. It is a prerequisite for responsible AI adoption.

True readiness exists when the five layers work in concert. When infrastructure, governance, and technical controls are aligned, AI can be deployed safely. When they are not, the organisation absorbs the consequences through reputational loss, regulatory breaches, financial exposure, and legal liability.

Network Security

Networks are the highways of the digital economy. Every email, payment, file transfer, and AI query travels across them.[18] They are fast, reliable, and often invisible, which makes them easy to take for granted. Yet like highways, they are exposed to accidents, congestion, and deliberate sabotage. Attackers can reroute traffic, intercept data, or quietly siphon valuable information. Networks can only be trusted when the right protections are in place.

Firewalls have long been a foundation of network security. They sit at the edge of the corporate environment, inspecting traffic and enforcing rules about who may enter and who must be kept out.[19] They remain a critical first line of defence, but the world in which they were designed to operate has changed. With cloud adoption and remote work, the perimeter has dissolved. Employees carry sensitive data on mobile devices, and applications run across multiple cloud services. Firewalls are still necessary, but they are no longer sufficient on their own. Additional layers such as intrusion detection, network segmentation, and VPNs are required to monitor traffic, reduce exposure, and secure remote access.

The shift toward Zero Trust reflects this new reality. No user or device is trusted by default, whether inside or outside the network.[20] Every access request must be verified, every session revalidated, and every identity confirmed. Technology, however, is only part of the answer. Success depends on leadership commitment, organisational discipline, and user awareness. Networks do not defend themselves. A single misconfiguration, a forgotten firewall rule, or an unmonitored port can compromise millions of dollars of investment.

In May 2021, Colonial Pipeline, the largest fuel pipeline operator in the United States, learned this the hard way.[21] Attackers gained access through a compromised VPN account that lacked multi-factor authentication. Within days, nearly half of the East Coast's fuel supply was disrupted. Petrol stations ran dry, prices spiked, and the U.S. government intervened. What began as one weak account became a national crisis.

Networks underpin critical infrastructure, from financial markets to healthcare and logistics. Their compromise can trigger cascading failures across entire economies. This is not an abstract IT issue; it is a fiduciary risk that requires oversight at the highest level.

Attacks against networks typically fall into two categories: passive and active.[23] Passive attacks involve quietly monitoring network traffic to capture information without changing it. Active attacks interfere by modifying data, impersonating users, or replaying old traffic to disrupt trust. Both are dangerous. A passive adversary may silently extract training data to build competing or malicious models, while an active one may inject poisoned data or manipulate traffic, degrading system reliability. These compromises often go undetected until their impact becomes irreversible.

Cryptography remains the strongest and most widely trusted method for protecting data in transit.[24] Encryption and digital signatures are essential to securing communication over untrusted networks, ensuring that data remains confidential and tamper-resistant during transfer. Every AI query, dataset exchange, and software update should be cryptographically protected, supported by sound key management and endpoint security.

Organizations must ensure that these protections are more than technical claims. Encryption standards such as TLS 1.3 should be applied consistently, IPsec must secure site-to-site traffic, and digital certificates should be properly managed and regularly rotated. These are not operational details but governance imperatives. They determine whether the organization's commitments to confidentiality, integrity, and authenticity are being upheld in practice.

An emerging concern is the impact of quantum computing on encryption standards.[25] The data transmitted today may need to remain confidential for decades. AI models trained on sensitive medical or financial datasets could retain their value for years. Adversaries who capture encrypted traffic now may simply be waiting until quantum computing advances make current algorithms breakable.

Strong authentication remains central to network trust.[26] Passwords are still the weakest link, often shared, reused, or poorly managed. Tokens, biometrics, and federated identity systems provide stronger protection by reducing reliance on passwords and making impersonation more difficult.[27] Tokens can be revoked quickly and limit access to short time frames. Biometrics verify identity through unique physical traits, while federated identity centralises secure verification across systems, improving both privacy and control.

Single Sign-On (SSO) extends this principle by allowing users to access multiple systems with one verified identity. When implemented securely, it improves user experience and reduces password fatigue, which in turn lowers the risk of credential reuse. However, SSO also concentrates access through a single point of failure. If an attacker compromises the identity provider, they can potentially access every connected system. This makes governance and monitoring of SSO configurations essential. Boards should confirm that SSO is integrated with multi-factor authentication, strong session controls, and anomaly detection to ensure that convenience does not come at the cost of security.

Directors should expect clarity on measurable indicators of network health. They should know the organization's average time to detect abnormal traffic, the number of unmonitored ports that remain open, and whether legacy protocols are still used for sensitive transactions. They should also understand how often encryption keys are rotated and how many high-risk exceptions are approved to bypass monitoring systems. These are not technical details. They are governance signals that show whether management has real visibility or is working on assumptions.

Regulation in this area is still developing. Governments are defining standards for network resilience and software transparency, and regulators are introducing penalties for insecure vendors or poor patching. Until these frameworks mature, boards must set expectations, demand evidence, and hold management accountable. Security is both a technical and a governance responsibility. It requires expertise to design and operate controls, and informed oversight to question, verify, and prioritize investment.

Customers trust that their data will be transmitted safely. Employees trust that systems will not expose them. Regulators trust that organizations will act responsibly. When that trust is broken, the effects are immediate and personal. A leaked medical record is a loss of dignity. A compromised payroll system affects real families.

Networks are critical infrastructure. Securing them requires encryption, authentication, monitoring, and collaboration between engineers, risk professionals, and leadership. Boards should not ask whether their networks are secure. They should ask whether the organization is resilient, able to withstand attack, recover quickly, and maintain trust when tested. The strength of that resilience defines both technical competence and leadership integrity.

Software Security

If networks are the highways, software is the vehicle. Every application, from payroll systems to customer portals to AI assistants, is made up of code. And every line of code is a potential vulnerability. Even the best engineers make mistakes, and attackers only need one.

The history of software vulnerabilities is sobering. In 2014, Heartbleed exposed a flaw in OpenSSL, a cryptographic library that underpinned much of the internet.[28] Attackers could read memory directly from servers, leaking passwords and encryption keys.

In 2017, WannaCry spread at unprecedented speed, exploiting unpatched Microsoft systems and crippling hospitals, banks, and shipping companies.[29] NotPetya followed, masquerading as ransomware but functioning as a state-backed attack that cost companies like Maersk and Merck billions.[30] In 2021, a flaw in the widely used log4j library became one of the most urgent software vulnerabilities in history.[31] Within hours of disclosure, attackers were scanning the internet for exposed systems. Because log4j was embedded everywhere, organizations scrambled to patch, some for months, some never fully succeeding.

Software vulnerabilities emerge daily. Attackers evolve faster than defenders, and patching delays can turn small flaws into crises. Governance in this space is not about eliminating risk but about measuring and managing it with discipline. Boards need clear visibility into how long it takes to patch critical vulnerabilities, how many outdated dependencies remain in the software stack, and how access controls enforce the principle of least privilege.

Software supply chains add further complexity. The 2025 breach at the Australian Human Rights Commission demonstrated how third-party components can become the weakest link.[32] Sensitive submissions from vulnerable individuals were exposed because of a flaw in a third-party form. The data was searchable online and potentially ingested into AI models scraping the web. This was not a sophisticated hack. It was a failure of vendor oversight with deeply human repercussions.

Governance lapses also explain why so many breaches are preventable. The incident in 2017, which exposed the personal data of 147 million Americans, was the result of a known flaw that went unpatched for months.[33] The vulnerability was public knowledge, but processes failed. The result was catastrophic loss of trust, financial penalties, and leadership change. Similarly, the SolarWinds compromise in 2020 highlighted how attackers can weaponize trust itself. By inserting malicious code into routine updates, they created a backdoor into thousands of organizations worldwide. SolarWinds' customers trusted the update process. Attackers turned that trust into a weapon.

Recent events show that AI has not escaped these dynamics. In July 2025, researchers discovered a critical vulnerability in Microsoft Copilot Enterprise, one of the flagship AI platforms being adopted by global corporations.[34] The flaw allowed unauthorized users to gain root access to Copilot's backend container, effectively granting them control over the environment in which the AI operated.

The issue began with an April update that introduced a new interactive coding feature designed to make Copilot more flexible for developers.[35] However, a small configuration error in the update allowed attackers to gain administrator-level access to the system that hosted the AI. In simple terms, the environment was built in a way that let an unauthorized user upload a file the system then trusted and executed with full privileges.[36] That single mistake provided complete control over the underlying software environment.

The exploit was deceptively simple. Yet in the context of an AI enterprise tool, its implications were profound. Root access meant attackers could manipulate how code executed, alter outputs, or explore the container file-system. Even though no sensitive data was found during the demonstration, the potential for misuse evident through tampered AI outputs, poisoned training data, or lateral movement into other Microsoft services.

These examples offer three lessons. First, AI systems are still software systems, built from the same code, libraries, and dependencies as any other application. Their risks are inherited from these components, as seen in the case of Jupyter Notebook and container orchestration scripts. Second, scale magnifies impact. A vulnerability in a widely distributed cloud-based AI tool can replicate instantly across thousands of corporate environments. Third, oversight gaps are costly. This was not a cutting-edge AI model flaw, but a textbook secure coding failure that would have been caught by stronger review and environment hardening.

Microsoft patched the issue by July 25 after responsible disclosure by Eye Security.[37] The flaw caused no damage beyond the demonstration, yet its existence shows how fragile trust can be. Boards that assume enterprise AI is inherently safe because it comes from a reputable vendor risk being blindsided. Safe adoption depends not only on what the model does but on the operational discipline that surrounds it.

This Copilot case reinforces the same point made by Heartbleed, log4j, and SolarWinds. Software security is a chain, and the chain breaks at its weakest link. In AI platforms, those links are not only models but the scripts, sandboxes, libraries, and integrations that make the model usable.

Software security is not simply a control. It is a commitment. Customers, employees, and regulators all trust organizations to act quickly, patch vulnerabilities, and manage dependencies responsibly. When that trust is broken, the consequences reach far beyond the technical. They damage reputation, legitimacy, and the social contract on which every organization depends..

Cloud Security

Cloud security refers to the practices, policies, and technologies that protect data, applications, and systems hosted in the cloud. Unlike traditional IT, where everything sat within a company's controlled data centre, cloud services shift infrastructure outside the physical perimeter. Providers such as Amazon Web Services, Microsoft Azure, and Google Cloud deliver immense speed and flexibility, offering elastic computing, global reach, and advanced services once out of reach for most organisations. Yet this flexibility introduces a new set of risks. Recognising how these risks intersect with emerging AI technologies is essential to safeguarding data and integrity in the cloud.

The first challenge is the shared responsibility model. Cloud providers secure the underlying infrastructure, including servers, storage, and networks. Organisations, however, remain accountable for how data is stored, who can access it, and whether configurations are secure. Many high-profile breaches have not resulted from flaws in the cloud itself but from customer-side misconfigurations. In 2019, Capital One suffered a breach that exposed the data of more than 100 million customers because an AWS S3 bucket was left misconfigured. The provider had met its obligations; the customer had not. Outsourcing infrastructure is not outsourcing accountability.

The second challenge is complexity. Few organisations operate in a single, neat environment. Hybrid models are now the norm. Some systems remain on-premises, others move to public or private clouds, and many spread workloads across multiple providers to avoid dependence on a single vendor. Each platform comes with its own dashboards, controls, and integration quirks. Stitching these together creates a patchwork that is difficult to govern.

Sensitive data may also reside in multiple jurisdictions, often without clear visibility into where it is stored or whether it complies with local regulations. The Australian Government has flagged this as a national security issue, noting that data sovereignty is not only about privacy but also about resilience in a crisis. Blind spots are governance gaps. Without oversight, it is impossible to know whether the organisation's most critical asset, its data, is being protected consistently.

A third challenge is identity and access management. In the cloud, identity is the new perimeter. Employees change roles, contractors cycle in and out, and accounts are often left active longer than necessary, creating "access creep." Shadow IT adds further risk, as staff adopt services such as Dropbox, Trello, or ChatGPT without approval, often storing data outside legal jurisdiction. Bring-your-own-device practices blur the boundary further, connecting personal devices to corporate systems. Each unmanaged account, unapproved tool, and insecure device expands the risk surface. Without proper controls, AI services can become another layer of exposure. A single employee pasting sensitive information into Copilot may inadvertently leak regulated data.

Addressing these risks requires a mindset shift. The old model of defending a fixed perimeter no longer works. In a borderless environment, the principle must be "never trust, always verify." Zero Trust should be embedded throughout. Every access request must be authenticated, every session validated, and every privilege limited to the minimum required. Multi-factor authentication should be standard. Dormant accounts should be disabled, and privileges reviewed regularly. Monitoring must be continuous. Misconfigurations, unusual behaviour, and the sudden appearance of new SaaS applications should be flagged in real time.

Boards should resist vague assurances such as "we patch regularly" or "we monitor our network." They should ask measurable questions such as:

• What is our average time to remediate critical vulnerabilities?
• How many privileged accounts remain active without review?
• What proportion of workloads are encrypted by default, both at rest and in transit?

Governance must also extend to contracts. Agreements with third-party cloud providers should explicitly address security, compliance, data sovereignty, incident reporting, responsibilities, and liabilities. It is tempting for organisations to focus on price and performance while leaving security obligations vague. This is where boards must be most vigilant.

Real-world breaches show the consequences when these controls fail. In 2020, Blackbaud, a cloud provider used by charities and universities worldwide, suffered a ransomware attack that exposed donor data across hundreds of institutions. In 2021, attackers exploited Microsoft Exchange's cloud email service, compromising tens of thousands of organisations. Misconfigured databases have repeatedly exposed health records, payroll data, and intellectual property. These are not mythical zero-days. They are governance failures caused by weak access controls, unmonitored configurations, and inadequate oversight.

Cloud risk is systemic. A single misconfiguration can scale across millions of records. A single compromised supplier can ripple through an entire sector. Unlike a traditional data centre breach that may be confined to one physical location, cloud failures are global by design. The very features that make the cloud attractive, such as scalability, reach, and integration, magnify the impact when things go wrong.

Cloud computing is also the foundation of modern AI. The immense computing power, storage capacity, and distributed infrastructure needed to train and operate advanced models exist only because of cloud platforms. This makes cloud security not only about protecting today's data but about safeguarding the digital foundations on which future intelligence depends. A passive adversary could quietly exfiltrate training data, while an active one might inject poisoned data or manipulate APIs. Both could exploit weaknesses that remain unnoticed until the damage is irreversible. If the cloud environment is insecure, then the AI built upon it is insecure.

Cloud delivers extraordinary opportunity, but it also amplifies risk. Security in the cloud is not a project with an endpoint. It is a continuous discipline of contracts, access controls, monitoring, and enforcement. Above all, it requires informed governance through collaboration between technical experts and leadership. Boards should understand enough to ask sharper questions, and management should be able to provide clear, verifiable assurance.

Reference

1. Yuri Diogenes and Dr. Erdal Ozkaya, Cybersecurity – Attack and Defense Strategies (3rd ed, Packt Publishing, 2022).
2. Ibid
3. National Institute of Standards and Technology (NIST), Framework for Improving Critical Infrastructure Cybersecurity (Version 1.1, 2018).
4. Clint Bodungen, ChatGPT for Cybersecurity Cookbook (Packt Publishing, 2024).
5. Mohammed Khalil, Supply Chain Attack Statistics 2025: Costs, Cases, Defenses (2025) https://deepstrike.io/blog/supply-chain-attack-statistics-2025
6. Omar Santos et. al, The AI Revolution in Networking, Cybersecurity, and Emerging Technologies (Packt, 2024).
7. National Institute of Standards and Technology (NIST), Artificial Intelligence Risk Management Framework (AI RMF 1.0) (US Department of Commerce, 2023)
8. Pascal Ackerman, Industrial Cybersecurity - Efficiently monitor the cybersecurity posture of your ICS environment (2nd Ed., Pack, 2021).
9. Ian Goodfellow, Yoshua Bengio and Aaron Courville, Deep Learning (MIT Press, 2016)
10. Omar Santos et. al, The AI Revolution in Networking, Cybersecurity, and Emerging Technologies (Packt, 2024).
11. Clint Bodungen, ChatGPT for Cybersecurity Cookbook (Packt Publishing, 2024).
12. Microsoft, Microsoft 365 Copilot Readiness (Microsoft Learn, 2025) https://learn.microsoft.com/en-us/microsoft-365/admin/activity-reports/microsoft-365-copilot-readiness?view=o365-worldwide.
13. Ibid.
14. Chris Miller, Chip War: The Fight for the World's Most Critical Technology (Scribner, 2022).
15. 'Data, Privacy, and Security for Microsoft 365 Copilot' (Web Page, Microsoft, 5 September 2025) https://learn.microsoft.com/en-us/copilot/microsoft-365/microsoft-365-copilot-privacy.
16. Petar Radanliev, The Rise of AI Agents: Integrating AI, Blockchain Technologies and Quantum Computing (Pearson, 2025).
17. Charles H Crawley, Cloud Penetration Testing (Packt Publishing, 2023).

18. Omar Santos, Samer Salam and Hazim Dahir, The AI Revolution in Networking, Cybersecurity, and Emerging Technologies (Pearson Education, 2024).
19. Chris McNab, Network Security Assessment (3rd ed, O'Reilly Media, 2017).
20. Nicholas DiCola and Anthony Roman, Microsoft Azure Network Security (Packt Publishing, 2023).
21. Ibid.
22. Jen Easterly, The Attack on Colonial Pipeline: What We've Learned & What We've Done Over the Past Two Years (CISA, 2023) https://www.cisa.gov/news-events/news/attack-colonial-pipeline-what-weve-learned-what-weve-done-over-past-two-years.
23. Chris McNab, Network Security Assessment (3rd ed, O'Reilly Media, 2017).
24. ISO/IEC 27001:2022 cls 6–10 (controls for confidentiality and integrity); DiCola & Roman (2023) ch 9 ('TLS and IPsec Best Practices').
25. Chris Miller, Chip War: The Fight for the World's Most Critical Technology (Scribner, 2022).
26. ISO/IEC 27001:2022 cls 6–10 (controls for confidentiality and integrity); DiCola & Roman (2023) ch 9 ('TLS and IPsec Best Practices').
27. Okta. "What Is Token-Based Authentication?" (January 23 2025). Available at: https://www.okta.com/identity-101/what-is-token-based-authentication/.
28. https://www.theguardian.com/technology/2014/apr/08/heartbleed-bug-puts-encryption-at-risk-for-hundreds-of-thousands-of-servers
29. Upguard, What is the WannaCry Ransomware Attack? https://www.upguard.com/blog/wannacry.
30. Andy Greenberg, 'The Untold Story of NotPetya, the Most Devastating Cyberattack in History' (Wired, 2018) https://www.wired.com/story/notpetya-cyberattack-ukraine-russia-code-crashed-the-world/.
31. Australian Signals Directorate, What is Log4 (2021) https://www.cyber.gov.au/about-us/advisories/2021-007-log4j-vulnerability-advice-and-mitigations.
32. Cassie Crossley, Software Supply Chain Security: Securing the End-to-End Supply Chain for Software, Firmware and Hardware (O'Reilly Media, 2024).
33. Australian Human Rights Commission, Data Breach Notification (2025) https://humanrights.gov.au/about/news/data-breach-notification.
34. Ravie Lakshmanan, 'Zero-Click AI Vulnerability Exposes Microsoft 365 Copilot

Data Without User Interaction' (The Hacker News, 2025) https://thehackernews.com/2025/06/zero-click-ai-vulnerability-exposes.html.

35. "Equifax Data Breach" (Electronic Privacy Information Center, 9 February 2020) https://archive.epic.org/privacy/data-breach/equifax/.
36. Guru Baran, 'Microsoft Copilot Rooted to Gain Unauthorized Root Access to its Backend System' (Cyber Security News, 2025) https://cybersecuritynews.com/microsoft-copilot-rooted/.
37. Ibid.
38. Erik van Klinken, 'Critical Vulnerability Discovered in Copilot Enterprise' (Techzine Europe, 2025) https://www.techzine.eu/news/security/133337/critical-vulnerability-discovered-in-copilot-enterprise/.
39. Glenn M Dotson, Practical Cloud Security (2nd ed, O'Reilly Media, 2024) ch 1 ('Cloud Security Fundamentals').
40. Shaharyar Khan, Ilya Kabanov, Yunke Hua and Stuart Madnick, 'A Systematic Analysis of the Capital One Data Breach: Critical Lessons Learned' (2023) 26 (1) ACM Transactions on Privacy and Security Article 3 https://doi.org/10.1145/3546068
41. Joe Tidy, 'Blackbaud Hack: More UK Universities Confirm Breach' (BBC News, 2020) https://www.bbc.com/news/technology-53528329.
42. Australian Cyber Security Centre, Advisory 2021-002: Active exploitation of vulnerable Microsoft Exchange servers (Web Page, 12 March 2021) https://www.cyber.gov.au/about-us/advisories/advisory-2021-002-active-exploitation-vulnerable-microsoft-exchange-servers
43. Amazon Web Services, Shared Responsibility Model (AWS Whitepaper, 2023).
44. Muhammad A Nomani, Mastering Cloud Security Posture Management (CSPM) (Packt Publishing, 2024).
45. Glenn M Dotson, Practical Cloud Security (2nd ed, O'Reilly Media, 2024).

"A chain is only as strong as its weakest link. In security, that link might be a server, a password, or an unlocked door."

CHAPTER 10

AI Security Risks

The foundation of security does not change with AI. Confidentiality, integrity, and availability remain the cornerstones.[1] Layered defenses across hardware, networks, software, and physical systems are still essential. Firewalls still matter. Encryption still matters. Identity still matters. What AI changes is not the principles, but the scale, speed, and autonomy of risk.[2]

AI systems operate at extraordinary speed. They process data in milliseconds, connect patterns across vast networks of information, and interact with countless systems at once. A human analyst might overlook a small anomaly, but an AI tool can detect and link weak signals automatically. A single neglected detail, such as an unprotected endpoint, an exposed API (Application Programming Interface), or a misconfigured identity control, can turn a minor flaw into a gateway for large-scale compromise.[3]

AI does not only consume data,[4] it generates it, transmits it, and in many cases acts on it. A search engine retrieves; an AI system interprets. A spreadsheet calculates; an AI agent executes. The difference is subtle until it isn't. AI links platforms together, triggers downstream actions, and in the case of autonomous agents, makes decisions without continuous human oversight. Every one of these interactions sits on infrastructure that may never have been designed for this level of interconnectedness.

That is why AI security cannot be treated as an add-on to cybersecurity.[5] The principles may be the same, but the context is transformed. APIs, agents, and misconfigurations are not new concepts. What is new, is the way AI magnifies them. The attack surface increases the ramification of errors and blurs the boundary between inside and outside threats.

Consider APIs. In a traditional application, an API might expose a small set of functions to a limited audience.[6] In an AI-driven enterprise, APIs become the lifeblood of workflows by feeding models with context, pulling data from multiple systems, and returning outputs that trigger the next link in a chain.

The risk is not simply that an API key is stolen. The risk is that a single key opens access to a model that has been entrusted with customer records, intellectual property, or financial transactions. In that environment, an exposed credential is not just a cost item on the IT ledger. It is an existential governance issue.

Or take AI agents. In a conventional system, execution flows are tightly scripted.[7] Developers define what a system can and cannot do. AI agents are different. They are given goals, not scripts. They decide which tools to call, which files to fetch, which APIs to invoke. That flexibility is their power, but it is also their danger. A poisoned prompt, a manipulated dataset, or even an ambiguously worded instruction can lead an agent to take actions no human ever intended. This is not speculation. Early experiments have already shown agents ex-filtrating secrets, overwriting files, and chaining together tools in ways that bypass ordinary controls. In the hands of adversaries, the same creativity that makes agents useful becomes a vector for exploitation.

Even the mundane becomes dangerous in this context. Misconfigurations have always been a bane of IT security. A storage bucket left public, a role assigned with too many privileges, a patch applied too late. These errors once created isolated incidents. In the age of AI, they create amplification loops. A misconfigured bucket connected to an AI pipeline may feed sensitive data into a model, where it is stored, indexed, or reproduced in outputs for months before anyone realizes. A weak identity control may not just give access to a single system; it may cascade through every workflow the model touches.

Boards should also recognize that AI accelerates the erosion of the boundary between internal and external threats. A misconfiguration in a developer sandbox can become an entry point to production. A prompt pasted into an unsanctioned AI tool can leak confidential strategy documents into external systems. The line between an accident and an exploit blurs when the same conditions enable both.

The governance challenge is not only about identifying new threats.[8] It is about understanding how existing weaknesses behave under new pressure. A leak in an ordinary plumbing system is inconvenient. The same leak in a high-pressure system can be catastrophic. AI is that pressure. It moves more data, faster, across more connections than any human process ever could.

Executives do not need to code an API or configure a cloud storage bucket. They need visibility and discipline. APIs must be mapped, their keys securely managed, and permissions for AI agents clearly defined. Cloud configurations must be monitored continuously, with alerts for drift and automatic remediation. Unsanctioned tools must be detected and removed before they undermine governance. These are the controls that turn AI from a black box into an accountable system.

The security story of AI is not about machines turning hostile. It is about humans underestimating how quickly their tools can be turned against them. The risks, including insecure APIs, over-permissive agents, misconfigurations, and exposed storage buckets, are not new.

Most of the measures to mitigate them already exist in the OWASP security frameworks, which set the baseline for modern API protection. What has changed is the scale, speed, and autonomy with which AI can amplify those weaknesses. Boards that treat AI security as a bolt-on will fall behind. Boards that treat it as governance, grounded in proven standards and continuous oversight, will stay ahead.

Types of AI Security Risks

APIs Risks for AI

APIs are unseen connectors of the digital economy. They allow systems to talk to each other, enabling everything from payments between banks to logistics updates between suppliers, or a chatbot pulling customer records into a conversation.[9] In an AI-enabled enterprise, APIs are everywhere. They make it possible for algorithms, platforms, and partners to exchange data and trigger actions seamlessly across the organization and beyond.

This same connectivity also creates exposure. Every API is both a bridge and a doorway.[10] It enables valuable transactions, but if left unsecured it can also provide a pathway for exploitation. Unlike interfaces designed for people, APIs are built for machines. They do not come with the guardrails humans expect such as login prompts, warnings, or obvious signals. To an attacker, or to an AI-driven exploit, a poorly secured API is not a locked door but a tunnel into the systems and data behind it.

Unlike traditional user interfaces, APIs are not designed for people. They are built for machines talking to machines. That means they lack the intuitive guardrails humans expect: login screens, warning pop-ups, visible alerts. To an attacker, an API endpoint is not a locked door but a direct tunnel into the systems and data behind it.

Insecure APIs are not always breached through advanced attacks. Many incidents come down to simple mistakes such as poor configuration, missing authentication, or an endpoint that was never closed. The 2022 Optus breach showed how damaging this can be.[11] An internal API, left open without basic protections, gave direct access to customer records through nothing more than a URL. There was no malware, no phishing, no sophisticated exploit. It was simply an unlocked door.

AI does not create new vulnerabilities, but it makes existing ones far more dangerous. Automated tools can scan entire networks, map endpoints, and generate traffic that appears legitimate.[12] What once depended on persistence or luck can now happen at machine speed, turning small oversights into major exposure.

The speed of automation matters. What once took weeks of manual probing can now be attempted in minutes. AI-driven tools can generate traffic that looks normal, making it harder for traditional detection systems to spot anomalies. This changes the scale of risk. A single weak API, left exposed, can be discovered and exploited far faster than ever before, turning what might once have been an isolated oversight into a large-scale breach.

Meta faced a related challenge. Its public endpoints allowed large-scale scraping of hashtags, profile data, and user posts.[13] This was not a hack in the traditional sense, but it exposed the fact that too much data was available, too quickly, without proper limits. Automated tools harvested information at industrial scale, far beyond what the system was designed to handle. Insecure APIs rarely look like attacks. They look like normal traffic, only faster, larger, and in the wrong hands.

The Open Worldwide Application Security Project (OWASP) has identified insecure APIs as one of the top ten risks in large language model applications.[14] This highlights how central APIs are to modern digital systems and why boards must pay attention. APIs are not peripheral to AI. They are its nervous system. Every AI tool that integrates with enterprise systems, whether a chatbot querying customer records or an analytics engine analyzing transactions, depends on APIs. If those APIs are insecure, the risks of AI adoption multiply.[15]

The danger comes from scale. A single misconfigured API can expose millions of records in minutes.[16] Privacy laws such as GDPR and Australia's Privacy Act impose heavy penalties for such breaches, and cyber insurers are increasingly refusing coverage when the lapse is as basic as missing authentication or weak key management. The reputational impact is just as serious. Customers do not distinguish between a flaw in an API and a flaw in the organization itself. To them, the breach is the brand.

Managing these risks requires discipline. Every API should require secure, token-based authentication, and anonymous access must be eliminated.[17] Rate limits and throttling should be in place to prevent scraping or brute force. All data in transit must be encrypted from end to end.[18] An unencrypted API call is no different from sending passwords on a postcard.[19] Monitoring and anomaly detection systems should be tuned to flag unusual behavior, whether by volume, pattern, or context. Old APIs must be decommissioned properly. A forgotten endpoint is an attacker's dream.

APIs in the age of AI place organizations at greater risk.[20] They expand the attack surface, link systems that were never designed to interact, and create openings that attackers can exploit at scale. As AI adoption accelerates, the number of API connections will only increase. Every chatbot, every analytics tool, every integration is another doorway that must be secured.

The governance responsibility is not to master the technical details but to ensure the organization has visibility, discipline, and accountability over API security. In an AI-driven enterprise, an insecure API is not a minor flaw. It is an open door. The task for boards is to make sure those doors are locked, monitored, and reinforced. Without getting too technical, lets look at some controls that could reduce this risk.

API Gateways[21]: An API gateway acts as a central checkpoint for all traffic. Instead of leaving every developer to build their own protections, the gateway enforces common rules across the enterprise. It controls authentication, encryption, throttling, and monitoring from a single place. This reduces inconsistency and ensures the board can have confidence that security is not fragmented. A gateway is not just a piece of technology. It is a governance tool that enforces discipline at scale.
Strong Key and Token Management[22]: APIs rely on digital keys and tokens to prove identity. If these are left hard-coded in code repositories, never rotated, or issued without expiry, they become open doors. Keys must be stored securely, rotated regularly, and tied only to specific applications. Poor key hygiene has been at the center of many breaches, often because it feels like a small detail until it is too late. Keys are the locks on digital doors, and weak locks invite compromise.

Secret Management[23]: APIs often rely on credentials such as passwords, certificates, or environment variables. These secrets can sometimes be mistakenly published in public repositories, logs, or screenshots. In 2025, security researchers reported a surge in automated scans targeting leaked Git configuration files, which often contain precisely these kinds of credentials. Strong practices, such automated scans of repositories, vault-based storage of secrets, and immediate rotation when exposures are detected, is essential.

Configuration Discipline[24]**:** Many API breaches stem from simple misconfigurations. The 2022 Optus breach is a prime example, where a forgotten API endpoint was left open without authentication or rate limits. Boards should expect the organization to carry out regular reviews of API settings, enforce minimum standards, and patch misconfigurations quickly.

Zero Trust Principles[25]**:** Zero Trust means never assuming a request is safe. Every user, device, or application must prove its identity and show it has only the minimum access required. Verification happens continuously, not just once. This way, if one system is compromised, the attacker cannot automatically move into others. The principle is simple: never trust by default, always verify. If an attacker finds a way into the network through a poorly secured API, they cannot move freely. Their device or service would be checked, and if it is not authorized, it would either be blocked immediately or isolated so it cannot spread further. The same principle applies whether the request comes from inside or outside the network. Every action must be verified before access is granted.

Access Control and Least Privilege[26]**:** Access must always follow the principle of least privilege. Every permission granted is a potential pathway to an exploit. APIs that process payroll, health data, or customer payments should be restricted to only the users and systems that need them. Boards should expect to see role-based access controls, regular reviews of who has access, and evidence that privileged accounts are tightly monitored.

Vendor Accountability and Contracts[27]: APIs often rely on third-party libraries, SDKs, and vendors, meaning any flaw in those components can flow directly into the organization. In 2024, a faulty update from CrowdStrike caused widespread system crashes across millions of devices. It was not a cyberattack but a configuration failure that revealed how fragile these dependencies can be, rippling across industries and disrupting operations globally. To reduce this risk, organizations must treat supply chain security as a priority. Components should be inventoried, scanned for vulnerabilities, and patched regularly. Contracts should also define what access is granted, how security standards are enforced, and what liability applies if those standards are breached.

Threat Modeling and Red Team Testing[28]: Threat modeling forces teams to think like attackers, asking how APIs could be exploited, what could go wrong, and what impact it would cause. Red team testing simulating real-world attacks with the same tools that adversaries use. These exercises identify gaps before criminals do. Boards should exports reports to include findings together with actions to close the gap.

Segmentation and Isolation[29]: Not all APIs carry the same risk. An API that displays a public catalogue is very different from one that processes payments or holds health records. If all APIs run on the same network, a weakness in the lowest-risk system can be used to reach the crown jewels. Segmentation fixes this by separating sensitive APIs into their own protected zones. These zones have tighter authentication, closer monitoring, and stricter controls. This makes lateral movement much harder for attackers and ensures that even if one API is compromised, others remain safe.

Data Minimization and Masking[30]: APIs should only share what is absolutely necessary. If full records are returned when only part of the data is needed, attackers gain more than they should. Sensitive fields should be masked, redacted, or removed before data leaves the system. Boards should confirm that data minimization is enforced as policy and practice. The less information exposed, the less damage if something goes wrong.

Continuous Discovery and Inventory[31]: The lack of API Inventory is a risk. According to Ofer Hakimi, "shadow APIs" are created and used without central oversight.[32] These type of APIs could remain open and unmonitored for years. Discovery tools can scan the environment to maintain a live inventory. Without this visibility, boards cannot be confident that what they are told represents the full picture.
Monitoring and Anomaly Detection[33]: API exploit can often look like ordinary traffic. Malicious actors may scrape data or flood requests in ways that resemble everyday use. Continuous monitoring helps reduce this exploit. Systems must be tuned to detect unusual behavior, whether in traffic volume, patterns, or context.
Audit Trails and Accountability[34]: Every API request should be logged in a way that cannot be altered. This creates an evidential trail of who accessed what, when, and from where. Logs are not just technical artifacts. They are governance safeguards that allow organizations to defend decisions in audits, inquiries, or court cases. Logs should be comprehensive, protected from tampering, and reviewed regularly.

AI Agents

In the race to harness AI's potential, a new class of tools known as AI agents has emerged. Unlike simple models that respond to prompts, these agents can plan, sequence, and execute tasks with minimal human oversight.[35] They can connect to APIs, browse the web, write and run code, and even interact with enterprise systems. Their promise lies in efficiency and automation at scale. Yet with that promise comes risk, amplified by the autonomy that makes them so useful.

Traditional software waits for explicit user instructions.[36] AI agents are different. Many can make limited decisions independently, chaining tasks together and acting on outputs as if they were new inputs. This partial autonomy creates opportunity but also exposure. A single prompt injection, such as a malicious input hidden in an email, webpage, or document, can cause an agent to execute unintended instructions.[37] If that agent has system-level access or elevated privileges, the consequences can be severe: deleting critical files, exfiltrating sensitive data, or installing malware at machine speed.

AI agents often rely on external connections to function effectively. They interface with APIs, databases, and third-party services to retrieve or act on information. If these connections are not properly secured, they can become high-value entry points. Credentials such as API keys or tokens are sometimes left in plain-text configuration files or public repositories without protection.[38] Attackers actively scan for these exposed secrets. If an agent is compromised, those credentials can provide unrestricted access to enterprise systems. In some cases, poorly configured agents are launched without authentication altogether, allowing anyone on the network—or even on the internet—to hijack them.

Recent incidents show how fragile this ecosystem can be. In September 2025, Salesforce patched a critical vulnerability called ForcedLeak.[39] Security researchers demonstrated that attackers could exploit the way AI-powered agents handled certain requests, forcing them to leak sensitive data without authorization. This was not a theoretical exercise but a real flaw in widely used enterprise software. It revealed that AI agents are already a target and that weaknesses in their design or integration can spread quickly across industries.

Plugins and extensions multiply this risk. Many agent frameworks allow external code modules to be added dynamically. While this expands capability, it also revives one of cybersecurity's oldest problems: running unverified code. A malicious plugin can act as a trojan horse, bypassing defenses because it operates through a trusted agent. The lack of robust logging and access controls in some agent frameworks compounds the problem, leaving organizations blind when things go wrong.

AI agents are not just IT experiments. They are potential gateways into critical systems.[40] An agent misused in customer support could generate fraudulent messages or harvest personal data. An agent connected to backend systems could leak trade secrets or financial records. Regulators will not accept "the AI did it" as an excuse. Liability remains with the business.

The speed of harm is a further risk. Human attackers typically probe systems step by step. A compromised AI agent can act continuously, across multiple systems, in real time. A single misconfiguration can escalate into a systemic breach before human responders even notice.

AI agents may boost productivity and unlock innovation, but without firm guardrails they represent an open door to the enterprise. The goal is not to master every command an agent can execute but to enforce policies, processes, and safeguards that prevent autonomy from turning into anarchy. In today's environment, the line between helper and hazard is thin, and waiting to recognize that distinction after the damage is done is too late.

Recent reporting, including BBC coverage of concerns about "rogue AI agents," highlights how quickly this risk is moving from theory to reality.[41] While AI agents can enhance efficiency and reduce costs, the essential question for boards is whether they can be introduced without exposing the business to unnecessary risk. The following controls outline how organizations can reduce the likelihood of agents behaving unpredictably or being exploited for harm.

Restrict and Review Agent Capabilities[42]: AI agents should be restricted to only the tools, functions, and access necessary for their defined purpose. Broad or open-ended permissions increase the risk of errors or unintended actions. Default settings must be reviewed and tested before release, as many systems include convenient but insecure configurations that expose data or bypass logging. After deployment, unused tools or extensions should be removed, and third-party access must be revoked once development ends. A clear off-boarding process ensures no residual access remains and reduces the likelihood of future compromise.

Controlled Deployment Rights: There should be clear rules about who is allowed to create and launch agents. Anyone with technical access could deploy an AI agent, including internal staff, external contractors, or junior developers. Only authorized personnel should deploy agents, and every deployment should be reviewed and recorded. This ensures traceability and accountability.

Least Privilege Access[43]: AI agents should only be given access to the systems and data they need. They should not have full access to everything. By allowing permissions to all systems, increases the risk of accidental or harmful behavior. Access should be limited by design, and reviewed regularly.

Agent Action Logging[44]: Logs of all AI Agents should be recorded and should include time, date, and details about what the agent did. Logs should be securely stored so that neither the AI agent nor unauthorized individuals can modify or delete them. Without proper logging, it becomes difficult to determine what went wrong when something fails.

Explainability by Design[45]: Where AI agents make decisions that impact people or carry significant risk, there must be a way to explain how and why those decisions were made. If an agent denies a loan application or flags a transaction, there should be a clear reason that can be checked. The Robodebt scheme showed what can happen when automated decisions are made without proper oversight or explanation. The result was significant consequences for government which is still being scrutinized.

Kill Switch[46]: There should be a method to shut down and AI Agent quickly, if something goes wrong. This includes both manual and automated controls which are activated if the agent behaves outside its expected limits. Boards should confirm that these safety measures exist and confirm who has the authority to use them.

Monitoring for Unexpected Behavior[47]: AI agents are built to learn and adjust over time. But without proper oversight, they can begin to act in ways no one expected. They might start making unusual decisions, accessing areas they shouldn't, or producing results that don't make sense. That's why ongoing monitoring is critical. It helps catch these changes early, so problems can be fixed before they lead to real damage.

Remove Unused or Legacy Extensions[48]: Sometimes tools or plugins used during development are left enabled even though they're no longer needed. These leftover components can introduce unnecessary risk. All unused extensions should be removed before deployment. This reduces the chance of unexpected agent behavior or exploitation through forgotten pathways.

Match Access to the Person: When an agent acts on behalf of a user, it should only have the same access that the user has. It should not be able to see more data or take actions beyond what the person is authorized to do. This helps maintain control and prevents overreach.

Check Vendor Contracts and Development Skills: Before engaging vendors or developers to build AI agents, organizations should ensure contracts are clear about security responsibilities, testing processes, and incident handling processes. It is also important to confirm that the people building the system have the right skills and understand the risks involved.

Misconfiguration

One of the most overlooked risks in AI deployment is misconfiguration.[49] Misconfiguration is where a system has been set up incorrectly or left with weak settings. Traditional tools like firewalls or endpoint protection often come with well-defined security defaults, AI systems are different.

They rarely arrive secure by default. The moment they are activated, they absorb the strengths (or weaknesses) of the environment around them.

Misconfiguration is not always obvious. It can emerge through forgotten legacy rules, poor policy settings, or well-meaning features that silently override earlier safeguards. Organizations may believe their systems are secure, only to discover that default settings have weakened protection. Dashboards may glow green, but the underlying controls may be fragile or incomplete.

An example of weak configuration is the rollout of Microsoft Copilot. In Microsoft's documentation it advises organizations to configure access controls, define content boundaries, apply retention policies, and activate data loss prevention settings before enabling the tool.[50] However, several organizations appear to enable Copilot without aligning it to internal policies or reviewing the implications for sensitive data handling.51

If role-based access is not tightly scoped, Copilot may surface information that a user technically has permission to access, but which was never intended to be retrieved, summarized, or presented out of context. This is not a flaw in the tool itself. It is a governance gap.

The 2025 Copilot incident illustrates how exposure lingers. Even after private GitHub repositories were locked down, cached versions remained accessible through Bing.[52] The issue was not that Copilot continued to pull data directly, but that earlier exposure had already been indexed and persisted. Once data is public, control may be lost long after the bucket or repository is secured.

AI tools are dynamic. They generate, act, and learn in real time. It can drive decisions that affect people, expose data, or distort outcomes. If retention rules are poorly applied, private information may appear in dashboards, reports, or summaries shared with the wrong users.

Boards should seek assurance before deployment. They need to understand if APIs are exposed and whether they are authenticated. They should ask how configuration is managed across cloud systems, SaaS tools, and AI models. They must understand what data is being retained, by whom, and under what conditions.

Another example of misconfiguration is the Capital One breach. In 2019 a misconfigured cloud storage bucket exposed sensitive data of more than 100 million customers.[52] The system was encrypted. However, access controls were not applied correctly. In the healthcare sector, similar missteps have allowed patient records, scans, and diagnostics to become publicly accessible, often through storage systems left open by accident.

What has changed with AI is scale, speed, and reach. APIs widen the surface of attack. AI agents blur the boundary between internal and external actors. And decisions made by one tool can propagate across platforms in ways that are hard to see and harder to unwind.

Boards should seek clarification of accountability. When something goes wrong, so does 'shifting blame'. Vendors claim their instructions were clear. Internal teams blame defaults. Regulators hold both sides responsible. Contracts with vendors should clearly define who owns configuration responsibilities and what safeguards are in place. Governance must come from within. Lets look at a few misconfiguration controls which could help business reduce this risk.

Clarify Configuration Governance: AI tools must be governed by a clear configuration model. This includes confirming who is responsible, which policies apply, and whether the tool is aligned to internal controls.

Boards should not assume vendors have applied secure defaults. Governance must be owned within the organization, not outsourced.

Configuration Baseline[53]: Every AI tool and connected API should have a configuration record. This should show what settings were changed, which defaults were kept, and the rationale behind those decisions. Without a clear baseline, there is no way to detect misalignment, drift, or non-compliance over time.

Secure-by-Design Onboarding: From the start, AI tools should be configured to use only what is needed. Minimum permissions, features, and APIs should be enabled. Anything unused must be blocked. Default access should never be assumed safe. Boards should confirm that access has been explicitly shaped for the intended purpose.

Independent Configuration Reviews[54]: Configuration should not be signed off by the same team that built the tool. Boards should require that AI systems undergo independent checks before going live. This includes verifying access, logging, API exposure, and output handling. A second set of eyes reduces blind spots and catches overlooked risk.

Vendor Contracts: Contracts must clearly define who owns configuration responsibility. Boards should ask whether vendors provide setup logs, attestations, and security guidance. If a tool is misused or breached due to weak configuration, the contract should leave no doubt about who is accountable and what safeguards were expected.

Drift Detection and Monitoring[55]: AI tools change over time. Settings that were safe yesterday may not be safe tomorrow. Boards should ask how configuration changes are tracked, what alerts are in place, and who is notified if a setting becomes risky. Continuous monitoring is essential to detect silent or accidental change.

Support Awareness: Configuration is often dismissed as a technical matter, but it carries legal, reputational, and operational repercussions. Boards should support briefings for risk, legal, and IT leaders using real-world examples like the Optus API breach or the NSW Education case to make the risks real and relatable.

Cloud Storage & Buckets

Cloud storage providers use "buckets" as containers for data.[56] They can hold everything from customer records and source code to financial reports and training datasets for AI. Think of them as digital filing cabinets in the cloud. If left unlocked or misconfigured, their contents become immediately visible to anyone on the internet.

The danger is not only theft. Exposed datasets can be copied, reused, or even incorporated into machine learning models without consent.[57] If training pipelines connect to unsecured buckets, sensitive information may be absorbed into AI systems in ways that cannot be reversed. Once data is integrated into a model, it is effectively permanent.

The 2025 Copilot incident illustrates how exposure can persist. Even after private GitHub repositories were locked down, cached versions remained accessible through Bing.[58] The problem was not that Copilot continued to pull the data directly, but that earlier exposure had already been indexed and stored. Once information is made public, control can be lost long after the original source is secured.

The risks from open buckets fall into three categories. Regulatory exposure is immediate, as many jurisdictions treat unencrypted or public personal data as a violation regardless of whether it was accessed. Under GDPR, fines can reach four percent of global turnover[59] and under Australia's amended Privacy Act penalties can climb in the millions. Operational exposure is just as severe, since open buckets often contain keys, tokens, or source code that can be leveraged for larger breaches. Finally, reputational exposure cannot be overstated. Customers rarely distinguish between a system being hacked and a system being left open. Either way, trust is broken.

There should be inventory of all buckets across providers, encryption of data by default, and clear review of access permissions. Automated scanning should be in place to detect when buckets are public or misconfigured, and "temporary" buckets should have enforced expiry dates to ensure they do not live on indefinitely. Vendors, too, must be held accountable, with contracts requiring them to meet equivalent security standards.

Measuring exposure is equally important. Boards should expect to see metrics, not vague reassurances. Examples include the percentage of buckets accounted for in a central register, the proportion of data encrypted at rest, the number of access anomalies detected in a quarter, and the average time it takes to close a misconfigured bucket. These numbers turn an invisible risk into something concrete and comparable against other exposures.

Managing the risk requires execution, not invention. Buckets should be private by default, with explicit approval required to make them public.[60]Data must be encrypted in transit and at rest. Access should follow the principle of least privilege so that no one has entry by default. Automated scanning tools need to be operational and continuously monitored. Lifecycle management must ensure that temporary buckets are truly temporary, with deletion policies enforced. Finally, misconfigurations should be built into incident response drills so that teams are prepared to react quickly when exposures occur.

Misconfigured buckets are more than a technical oversight. They are a governance failure that can cascade into regulatory fines, operational breaches, reputational damage, or even biased AI models. Once data is exposed to the cloud, and once AI touches that cloud, the risks spread across the enterprise. Let's reviwe several controls which can help reduce this risk.

Default Privacy: Buckets should be private by default, with public exposure requiring explicit approval. This shifts the burden of proof onto those requesting openness, ensuring risk is considered upfront rather than patched later.

Encryption[61]: Data, both at rest and in transit, should be encrypted. Regulators increasingly treat unencrypted data as negligence, even if it is never breached. Encryption demonstrates diligence and strengthens regulatory defense.

Role-Based Access: Access must follow least privilege principles. Engineers, contractors, or applications should only have rights to the specific data they need. This reduces both accidental leaks and deliberate misuse.

Automated Monitoring[62]: Tools should continuously scan for exposed or misconfigured buckets. The speed of discovery is critical: exposure often becomes exploitation within minutes. Boards should expect metrics on how quickly misconfigurations are detected and closed.

Lifecycle Management: Misconfiguration often comes from neglect rather than malice. A "temporary" storage bucket created for a project can sit open for years if no one owns it. Lifecycle management sets rules to prevent this drift. Buckets should have defined expiry dates, mandatory reviews, and enforced deletion policies.

Testing and Simulation[63]: Exposure scenarios should be built into tabletop exercises. Boards need assurance that if an open bucket is discovered, the organization can respond quickly, notify regulators, and control reputational fallout.

Third-Party Accountability and Contracts: Many buckets are hosted and managed by third parties. Third Party mistakes become an organization's exposure. Strong contracts ensure vendors maintain security baselines such as encryption, access monitoring, and continuous scanning. Agreements should also define liability and notification obligations, so vendors cannot quietly pass the risk back to you. This is not lifecycle discipline inside the enterprise. This is holding external partners accountable to the same standards.

Insecure Output Handling

OWASP defines Insecure Output Handling as the failure to validate, sanitize, or control outputs generated by a large language model (LLM) before passing them into downstream systems.[64] In other words, when an AI response is trusted too freely, it becomes a new pathway for exploitation.

What does that look like in practice? Suppose an AI chatbot issues a command or a piece of code, and the system simply runs it because it "came from" the AI. Or imagine the model generates HTML, JavaScript, or SQL content that is displayed or executed without escaping or checking. Through those gaps, attackers can perform cross-site scripting (XSS), SQL injection, or remote code execution.

For example, a summarization tool powered by an LLM may fail to encode or escape its output before inserting it into a web page. A malicious user could craft a prompt that causes the model to produce harmful JavaScript, which then runs in the browser of every viewer, stealing session cookies, redirecting pages, or issuing unauthorized actions. Similarly, backend services that accept and run LLM-generated code without sandboxing can open the door to command execution.

The OWASP Top 10 for LLMs lists insecure output handling among the most serious vulnerabilities.[65] Empirical research supports this, showing that a significant proportion of code produced by LLMs contains exploitable flaws, including XSS and injection vulnerabilities.[66] When organizations deploy AI systems that act on model outputs without validation, they inherit these weaknesses by design.

The risk is that LLM outputs are shaped by both prompts and training data. A well-crafted input, known as a prompt injection, can trick the model into generating unsafe or misleading content that bypasses normal safeguards. When that output is then acted on without validation, the weakness shifts from input manipulation to insecure output handling. In this way, attackers can turn the model's trusted responses into a bridge to backend systems or higher privileges. Let's review several controls that could reduce this risk.

Zero-Trust Treatment[64]: Model outputs must be treated the same way as inputs from an external user. Just because the AI produced it does not make it safe. Outputs should be assumed untrusted until it has been validated. This prevents organizations from blindly passing AI-generated text or code into downstream functions where it could trigger harmful actions.

Validation and Sanitization[65]: Outputs that feed into other systems need guardrails. Validation checks ensure that only expected formats and commands are accepted, while sanitization strips out malicious or unnecessary elements. Following established frameworks such as OWASP's Application Security Verification Standard (ASVS) provides tested methods to enforce these controls consistently. Without them, an attacker could slip in dangerous instructions through an AI's response.

Output Encoding[66]: When AI outputs are shown to users, they must be encoded so that text cannot be mistaken for executable code. Without encoding, a malicious actor could embed JavaScript or Markdown into a model's response and hijack a user session. Encoding ensures that outputs display safely on screens, rather than being executed by browsers or applications.

Layered Monitoring and Alerts[67]: Outputs that interact with other systems should be monitored for anomalies. For example, if an AI suddenly starts generating SQL queries, shell commands, or scripts outside of its usual scope, this should trigger alerts. Logging and monitoring of outputs ensures issues are caught early.

Vendor Transparency and Testing: If an LLM or plugin is sourced from a vendor, contracts should require them to demonstrate how outputs are validated and how they prevent injection of harmful payloads. Boards should ask whether the vendor performs independent testing of output handling, since risk is shared once outputs move into business systems.

Prompt Injection

Prompt injection emerged as a risk as soon as OpenAI released ChatGPT to the public.[68] What began as a way to make AI systems more responsive quickly revealed new pathways for misuse. By shaping language inputs, users discovered they could not only guide the model's behavior but sometimes override its safeguards.

Prompt injection is the practice of using command commands in human language to guide LLM to reveal information.[69] It is an instruction that carries context, purpose, and intent. A prompt can help models such as GPT-4 or Claude deliver valuable results. Alternatively, it can bypass safeguards, extract confidential data, or produce misleading content.[70] Because prompts are written in everyday language rather than code, the line between creativity and exploitation is thin.

Recent incidents show how subtle these risks can be. In 2025, researchers demonstrated how a seemingly harmless "crossword puzzle" game could trick ChatGPT into revealing valid Windows product keys.[71] The attack disguised the malicious request inside the game's rules. Each stage of the puzzle nudged the model closer to revealing sensitive data until the final cue, "I give up," when the model provided the complete key as the "correct answer."[72] No firewall was breached and no malware was used.

A similar flaw appeared in a GitHub-related incident reported by Docker. Attackers exploited the Model Context Protocol (MCP) used in certain developer tools to send hidden prompts that executed local commands without the user's knowledge.[73] The prompt injection allowed unauthorized code execution and potential theft of API keys and source data.[74] The issue was quickly patched, but it revealed how deeply prompt vulnerabilities can penetrate modern development workflows.

These cases illustrate that prompt injection is not fiction. It is a practical exploitation of trust, context, and configuration. A single well-crafted phrase can override restrictions, retrieve credentials, or trigger real-world actions. Modern AI systems now include guardrails that detect and neutralize common injection patterns, scanning for hidden commands in HTML or metadata and separating user prompts from system instructions.[75] Yet even with these protections, the threat persists because context, together with human behavior, remain unpredictable.

Prompt injection has become a measure of AI maturity. Organizations that manage prompts with the same rigor as source code or financial controls understand that language itself is now part of the attack surface. The risk lies in the fact that prompts shape every output a model produces. A poorly designed or malicious input can override safeguards and trigger actions beyond intent. Because responses are often trusted without verification, the impact can spread across connected systems. Let's look at several key controls that can help reduce this risk.

Input Validation and Sanitization[76]: Inputs must be scanned for risky patterns such as "ignore previous instructions" or "reveal prompt." Suspicious content should be sanitized or removed before reaching the model. Following OWASP's Prompt Injection Prevention guidance ensures these defenses are applied consistently.

Output Monitoring and Validation[77]: AI outputs should be checked for signs of exploitation such as leaked configuration data or executable commands. Suspicious results must be filtered or blocked before reaching downstream systems.

Human-in-the-Loop (HITL): High-risk prompts containing terms like "password," "admin," or "override" should trigger manual review. HITL controls prevent sensitive actions from being executed automatically and should be reserved for high-impact operations.

Best-of-N Attack Mitigation[78]: Attackers often test multiple prompt variants to find one that succeeds. Rate limiting, circuit breakers, and behavior-based monitoring slow these attempts and giving time to respond.

Secure Implementation Pipeline and Testing[79]: A secure pipeline combines input filtering, structured prompts, output validation, and a kill switch. Regular red-teaming and adversarial testing expose weaknesses early and keep defenses current as models evolve.

Training and Awareness: Staff must understand how prompts influence outcomes, what data is safe to use, and how misuse creates risk. Training should cover data handling, use of approved tools, and persistence of information in model logs. Awareness must be ongoing so users treat AI interfaces as controlled entry points, not open text boxes.

Contracts and Agreements: Vendor and API contracts must clearly define data handling, storage, deletion, and incident response. Agreements should address prompt injection risk, audit rights, and obligations to follow frameworks such as OWASP LLM guidance or ISO/IEC 27001. Precise terms prevent disputes over ownership, liability, and remediation.

Acceptable Use Policy: This sets boundaries for acceptable AI use and clarify accountability within the organization. They must state that outputs require verification, sensitive data cannot be entered into public models, and adherence to company policy is mandatory. Enforced consistently, these agreements link governance policy to daily practice.

Security in artificial intelligence is not achieved through a single control but through the discipline applied across all of them. Each safeguard, whether protecting APIs, filtering outputs, or defending against prompt manipulation, serves a common purpose to preserve the integrity of human decision-making in an increasingly automated world. When directors ensure that these mechanisms work together, they are not only reducing technical risk but also strengthening trust.

Reference

1. ISO/IEC 27001:2022 – Information Security, Cybersecurity and Privacy Protection – Information Security Management Systems (International Organization for Standardization, 2022).
2. Gregory J Falco and Eric Rosenbach, Confronting Cyber Risk: An Embedded Endurance Strategy for Cybersecurity (Oxford University Press, 2022).
3. K Scott Griffith, The Leader's Guide to Managing Risk (Harvard Business Review Press, 2014).
4. Paul Daugherty and H James Wilson, Radically Human: How New Technology Is Transforming Business and Shaping Our Future (Harvard Business Review Press, 2022).
5. Microsoft, Fostering Appropriate Reliance on GenAI — Overreliance on AI: Risk Identification and Mitigation Framework (Microsoft Learn, 4 March 2025) https://learn.microsoft.com/en-us/ai/playbook/technology-guidance/overreliance-on-ai/overreliance-on-ai.
6. OWASP, API Security (2025) https://owasp.org/www-project-api-security/.
7. Reid Blackman, 'Organizations Aren't Ready for the Risks of Agentic AI' (2025) Harvard Business Review https://hbr.org/2025/06/organizations-arent-ready-for-the-risks-of-agentic-ai.
8. Douglas W Hubbard and Richard Seiersen, How to Measure Anything in Cybersecurity Risk (John Wiley & Sons, 2016).
9. OWASP, API Security Project (Open Worldwide Application Security Project, 2024) https://owasp.org/www-project-api-security/.
10. Ibid.
11. Optus Breach News Article (specific source to be cited when finalized).
12. Maurício Harley, Pentesting APIs: A Practical Guide to Discovering, Fingerprinting and Exploiting APIs (Packt Publishing, 2024) ch 8 ('Data Exfiltration and Enumeration Attacks').
13. Mehdi Medjaoui et al, Continuous API Management: Making the Right Decisions in an Evolving Landscape (2nd ed, O'Reilly Media, 2022).
14. OWASP Foundation, OWASP API Security Top 10 (2023) https://owasp.org/www-project-api-security/.
15. Maurício Harley, Pentesting APIs (Packt Publishing, 2024).
16. Advanced API Security (2020) ch 4; Pentesting APIs (2024) ch 8 (data exposure and leakage).

17. Prabath Siriwardena, Advanced API Security: OAuth 2.0 and Beyond (2nd ed, Apress, 2020) chs 2–4 (OAuth 2.0 flows and key management); OWASP Foundation, OWASP API Security Top 10 (2023) https://owasp.org/www-project-api-security/.
18. Advanced API Security (2020) ch 3 (TLS and Transport Layer Security controls); Mehdi Medjaoui et al, Continuous API Management (2nd ed, O'Reilly Media, 2022) ch 4 (Security pillar).
19. Ibid.
20. Olaf Zimmermann et al, Patterns for API Design: Simplifying Integration with Loosely Coupled Message Exchanges (Pearson Addison-Wesley, 2023).
21. OWASP Foundation, OWASP API Security Top 10 (2023) https://owasp.org/www-project-api-security/.
22. Prabath Siriwardena, Advanced API Security: OAuth 2.0 and Beyond (2nd ed, Apress, 2020).
23. Maurício Harley, Pentesting APIs (Packt Publishing, 2024).
24. OWASP Foundation, OWASP API Security Top 10 (2023) https://owasp.org/www-project-api-security/.
25. Ibid.
26. Ibid.
27. Mehdi Medjaoui et al, Continuous API Management (2nd ed, O'Reilly Media, 2022).
28. OWASP Foundation, OWASP API Security Top 10 (2023) https://owasp.org/www-project-api-security/.
29. Ibid.
30. Ibid.
31. Mehdi Medjaoui et al, Continuous API Management (2nd ed, O'Reilly Media, 2022).
32. Ofer Hakimi, Shadow APIs: Understanding the Risk and 6 Ways to Reduce It (2025) https://www.pynt.io/learning-hub/api-security-guide/shadow-apis-understanding-the-risk-and-6-ways-to-reduce-it
33. OWASP Foundation, OWASP API Security Top 10 (2023) https://owasp.org/www-project-api-security/.
34. Douglas W Hubbard and Richard Seiersen, How to Measure Anything in Cybersecurity Risk (2nd ed, Wiley, 2023).
35. Petar Radanliev, The Rise of AI Agents: Integrating AI, Blockchain Technologies, and Quantum Computing (Pearson, 2025).
36. Michael P Wellman, Trading Agents (Morgan & Claypool, 2011) 1–5 (agent decision logic and automation).

37. OWASP Foundation, OWASP API Security Top 10 (2023) https://owasp.org/www-project-api-security/.
38. Ibid.
39. Ravie Lakshmanan, 'Salesforce Patches Critical ForcedLeak Bug Exposing CRM Data via AI Prompt Injection' (2025) The Hacker News https://thehackernews.com/2025/09/salesforce-patches-critical-forcedleak.html.
40. Petar Radanliev, The Rise of AI Agents (Pearson, 2025).
41. Sean McManus, 'How to Stop AI Agents Going Rogue' (2025) BBC News https://www.bbc.com/news/articles/cq87e0dwj25o.
42. OWASP Foundation, OWASP API Security Top 10 (2023) https://owasp.org/www-project-api-security/.
43. Ibid.
44. Ibid.
45. European Union, Artificial Intelligence Act (Reg (EU) 2024/1689) arts 9–16.
46. OWASP Foundation, OWASP API Security Top 10 (2023) https://owasp.org/www-project-api-security/.
47. Ibid.
48. Ibid.
49. Muhammad A Nomani, Mastering Cloud Security Posture Management (CSPM) (Packt Publishing, 2024).
50. Microsoft Corporation, Data, Privacy, and Security for Microsoft 365 Copilot (Web Page, 5 September 2025) https://learn.microsoft.com/en-us/copilot/microsoft-365/microsoft-365-copilot-privacy
51. Miroslav Milovanovic, 'Microsoft Copilot Data Security and Governance: A Practical Guide for CISOs' (Knostic, 6 June 2025) https://www.knostic.ai/blog/microsoft-copilot-data-security-governance
52. Josh Taylor, 'NSW Education Department Caught Unaware after Microsoft Teams Began Collecting Students' Biometric Data' (2025) The Guardian https://www.theguardian.com/australia-news/2025/may/19/nsw-education-department-caught-unaware-after-microsoft-teams-began-collecting-students-biometric-data.
53. Capital One, Information on the Capital One Cyber Incident (2022) https://www.capitalone.com/digital/facts2019/.
54. Muhammad A Nomani, Mastering Cloud Security Posture Management (CSPM) (Packt Publishing, 2024).
55. OWASP Foundation, OWASP Security Misconfiguration https://owasp.org/Top10/A05_2021-Security_Misconfiguration/.

56. Rieke Müller, Mohamed Abdelaal and Davor Stjelja, 'Open-Source Drift Detection Tools in Action: Insights from Two Use Cases' (2024) Proceedings of the ACM Conference (Conference'17) (Association for Computing Machinery).
57. Glenn M Dotson, Practical Cloud Security (2nd ed, O'Reilly Media, 2024).
58. Petar Radanliev, The Rise of AI Agents (Pearson, 2025).
59. Ziv Karliner, 'New Vulnerability in GitHub Copilot and Cursor: How Hackers Can Weaponize Code Agents' (2025) Pillar Security Blog https://www.pillar.security/blog/new-vulnerability-in-github-copilot-and-cursor-how-hackers-can-weaponize-code-agents.
60. Intersoft Consulting Services AG (n.d.). General Data Protection Regulation (GDPR) – Legal Text. Retrieved October 30, 2025, from https://gdpr-info.eu/ Glenn M Dotson, Practical Cloud Security (2nd ed, O'Reilly Media, 2024).
61. Muhammad A Nomani, Mastering Cloud Security Posture Management (CSPM) (Packt Publishing, 2024).
62. Charles Horton Crawley, Cloud Penetration Testing (Packt Publishing, 2023).
63. Glenn M Dotson, Practical Cloud Security (2nd ed, O'Reilly Media, 2024).
64. OWASP, OWASP Top 10 for Large Language Model Applications (2025) https://owasp.org/www-project-top-10-for-large-language-model-applications/.
65. Ibid.
66. Chen et al, 'Do Large Language Models Produce Secure Code? A Study on XSS and Injection Vulnerabilities' (arXiv preprint, 2023) https://arxiv.org/abs/2310.02059
67. OWASP, OWASP Top 10 for Large Language Model Applications (2025) https://owasp.org/www-project-top-10-for-large-language-model-applications/
68. Ibid.
69. Douglas W Hubbard and Richard Seiersen, How to Measure Anything in Cybersecurity Risk (2nd ed, Wiley, 2023)
70. OpenAI, Introducing ChatGPT (30 November 2022) https://openai.com/blog/chatgpt
71. Gilbert Mizrahi, Unlocking the Secrets of Prompt Engineering (Packt Publishing, 2023)
72. Ibid.
73. NSFOCUS Global. Prompt Word Injection: An Analysis of Recent LLM Security Incidents (2025).
74. Ibid

75. Docker Blog. MCP Horror Stories: GitHub Prompt Injection (2025) (https://www.docker.com/blog/mcp-horror-stories-github-prompt-injection/)
76. Ibid.
77. OWASP, LLM Prompt Injection Prevention (2025) https://cheatsheetseries.owasp.org/cheatsheets/LLM_Prompt_Injection_Prevention_Cheat_Sheet.html
78. Ibid
79. Ibid

EXECUTIVE SUMMARY (Part2)

Risk is not new to business, but AI changes its nature. It amplifies existing vulnerabilities, accelerates their impact, and compresses the time between cause and consequence.

This section began with the fundamentals: understanding likelihood and consequence, and how inherent and residual risk define exposure. It showed that while AI introduces new threats, the framework for managing them remains grounded in established governance and risk practices.

From there, the focus turned to AI-specific risks such as bias, data poisoning, prompt injection, and model drift, as well as softer but equally damaging failures like over-reliance, loss of human judgment, and strategic drift. These risks expand as systems grow more autonomous and interconnected.

The discussion then moved to security, identifying the Big Four foundations that protect every AI environment: organizational readiness, network security, cloud security, and software security. Each pillar reinforces the others, creating the conditions for resilience.

Finally, the section explored the controls and mitigants that turn awareness into assurance, from explainability testing and data governance to model monitoring and human oversight. Together they form the guardrails that prevent vulnerabilities from becoming liabilities.

AI security is not about eliminating every threat. It is about embedding a culture of visibility, discipline, and shared responsibility. Trust in AI depends on trust in the systems and people that govern it.

KEY TAKE AWAYS

- The fundamentals of risk were discussed, illustrating how tools such as the risk matrix, likelihood and consequence analysis, and the concepts of inherent and residual risk can be applied to assess an evolving AI environment.
- AI-specific risks were explored such as Shadow AI, Black Box systems, supply chain exploits, model drift, and over-reliance.
- The Big Four were introduced as the foundations of AI security. Together, they form the fundamentals of the AI security landscape.
- AI security risks were also explored, including APIs, AI agents, misconfiguration, and cloud or bucket exposures, along with the mitigants designed to reduce those risks.

PART 3

Governance, Ethics and Law

"Boards do not have the luxury of curiosity without caution. Fiduciary duty demands both Imagination and Restraint."

CHAPTER 11

Accountability in the Age of AI

Fiduciary duty is not a new concept.[1] For more than a century, corporate law has required directors and officers to exercise care, loyalty, and good faith in guiding the companies they serve.[2] These duties have always been dynamic, stretching to cover new contexts as business models evolved. When railroads reshaped global commerce in the 19th century, directors were expected to understand not just finance but also operational risk on a scale previously unimaginable.[3] When electrification and heavy industry defined the early 20th century, fiduciary oversight expanded to include worker safety, industrial standards, and large-scale capital investments.

The digital era added another layer. In the late 20th century, as corporations became increasingly dependent on software and the internet, fiduciary duties grew to encompass oversight of IT, cybersecurity, and data privacy.[4] The failure of boards to anticipate these risks often led to shareholder lawsuits, regulatory penalties, and reputational collapse. Each wave of innovation has carried the same message: directors do not must understand enough to ask the right questions, seek the right advice, and make informed decisions.

Artificial intelligence represents the next frontier. What distinguishes AI is not just its power to automate tasks, but its ability to influence judgment by filtering data, shaping decisions, and interacting directly with customers, regulators, and employees. When algorithms drive lending approvals, medical diagnoses, hiring decisions, or strategic forecasting, fiduciary duties are no longer confined to financial statements.[5] They extend into how machines shape the organization's choices, and by extension, its liabilities.

This does not rewrite the law. The duties of care, loyalty, and good faith remain unchanged. What shifts is the standard of application. Just as courts once judged directors on whether they had interrogated financial assumptions or supply-chain exposures, they may now assess whether boards asked the right questions about algorithmic bias, data provenance, cybersecurity safeguards, or vendor accountability. Fiduciary duty in the age of AI is not a new rule. It is the old rule meeting a new reality.

Fiduciary Duties in the Age of AI

Corporate law has always imposed clear and enforceable duties on directors and officers. In Australia, the *Corporations Act* 2001 (Cth) sets out a suite of obligations that remain the foundation of governance in both stable and turbulent times. Among the most critical are the duty to act with care and diligence (s180), the duty to act in good faith and in the best interests of the corporation (s181), and the duties to avoid improper use of position (s182) or information (s183).[6] Collectively, these provisions require directors to behave as a reasonable person would in the same role, factoring in the nature of the company, their responsibilities, and the circumstances of the decision. They are not technical rules to be applied narrowly; they are broad obligations designed to hold leaders to a standard of judgment and stewardship that protects both the company and the community it serves.

These sections, while central, are only a sliver of the broader corporate law landscape.[7] Governance covers everything from equity finance to creditor protections, dividend distributions, takeovers, and voluntary administration. Each of these areas carries weight and complexity, but in the context of artificial intelligence, data, and digital risk, there is one theme that consistently rises above the rest: the standard duty of care. The technological environment that companies now inhabit is fast-moving, unpredictable, and profoundly consequential. Emerging tools demand that directors act with a level of diligence and foresight that is no longer optional. Understanding the systems being adopted, interrogating their risks, and governing their use is not just prudent leadership. It is a legal expectation.

In practice, these obligations are often described in terms of the duty of care and the duty of loyalty.[8] The duty of care requires directors to properly inform themselves, to understand the technologies their organizations adopt, and to weigh the risks they introduce.[9] The duty of loyalty requires that decisions, including those involving artificial intelligence, are made in good faith, for a proper purpose, and in the best interests of the company rather than for personal or short-term gain.

These principles are centuries old, rooted in trust law and fiduciary responsibility, yet they are finding new relevance in the digital age. Increasingly, regulators and courts are extending these expectations into the realm of technology governance, treating failures in oversight of digital systems as breaches of fiduciary duty.

Good governance standards explain what this looks like in practice. Acting in good faith for a proper purpose.[10] Avoiding material personal interests. Informing oneself appropriately about the issues at hand. Genuinely believing that decisions are in the best interests of the company. These are not abstract ideals. They are standards against which directors may be judged in court.

When applied to artificial intelligence, they mean that boards cannot simply approve AI initiatives without scrutiny. Directors must understand how these systems work, where their limitations lie, and what risks they pose to customers, employees, and the wider community. Ethical leadership and fiduciary responsibility converge at this point: to act diligently in the age of AI is to treat technology as a matter of governance, not just operations.

This has practical consequences. The law does not require directors to understand tokenization, training parameters, or the intricacies of reinforcement learning. But the law does require them to ask the right questions, seek the right advice, and maintain active oversight of emerging risks. A boardroom that passively observes technological change has already fallen short. Fiduciary duties demand an informed and proactive stance.

Duty of Care extends to digital governance. Oversight of AI systems, cybersecurity, and data privacy is no longer optional. Algorithmic bias, misuse of personal data, opaque decision-making, and systemic vulnerabilities are risks that boards must treat as foreseeable and therefore governable. Failure to do so may result in personal liability.

In *ASIC v RI Advice Group Pty Ltd* (2022), the Federal Court found that neglecting foreseeable cyber risks amounted to a breach of directors' duties.[11] The case illustrated that digital risk is not separate from traditional governance. It is central to it.

Case law further reinforces this trajectory. In *ASIC v Adler* (2002), misuse of position underscored the importance of transparency and accountability, principles that apply equally when boards are navigating AI.[12] In the Centro case (*ASIC v Healey*, 2011), directors were held liable for failing to notice basic errors in financial statements.[13]

Directors cannot delegate responsibility blindly, even when relying on management or external advisors. If they fail to interrogate the information before them, they breach their duties. If boards rely too heavily on AI insights or let hidden systems make decisions without checking them, they could fail in their duties. Even contemporary technology can fold, such as with FTX.[14] Although it is not an AI case, these events highlights negligent oversight in complex, technology-driven businesses.

Directors who fail to question investments in algorithm-driven systems, or who chase speed and hype over substance, risk breaching their duty of care. Regulators are unlikely to accept "we didn't understand the technology" as a defense. Boards must ensure that innovation is pursued with diligence, accountability, and a focus on long-term stability.

AI systems are increasingly intergrated into the core operations of companies. They filter CVs, approve loans, detect fraud, optimize supply chains, and shape customer interactions.[15] These are not peripheral functions. They are mission-critical processes that touch customers, employees, and regulators directly. That means the director's duty of care must be applied with the same rigor as financial oversight. Boards are expected to evaluate AI tools with attention to risk, transparency, and accountability. They must confirm that appropriate safeguards are in place, that governance frameworks are clear, and that adoption aligns with the company's long-term interests as well as its societal responsibilities.

The rise of AI has not changed fiduciary duty. It has sharpened it. Duties that were once centered on financial prudence and loyalty now extend to data governance, cybersecurity, and technological oversight. The law continues to evolve, but its trajectory is explicity. Boards are accountable for the consequences of digital transformation, whether or not they possess technical expertise.

The role of the board in AI strategy and oversight must therefore be re-imagined.[16] Artificial intelligence is not just an operational tool; it is a strategic enabler that affects customers, regulators, investors, and society at large. That makes it a board-level concern. Directors cannot abdicate responsibility to CIOs, data scientists, or third-party vendors. While operational management rests with executives, material decisions that carry financial, reputational, or regulatory risk demand board visibility and oversight.

Good governance begins with clarity. Boards must define ownership of AI risks and decisions through accountability frameworks such as RACI models, governance committees, or independent oversight panels. Just as importantly, they set the tone from the top.[17] When directors frame AI adoption as part of corporate values and long-term purpose, management follows. When they treat it as a race or a checkbox, shortcuts occur.

The board's duty extends beyond approving initiatives. It includes ensuring that guardrails prevent AI from drifting into unethical or unlawful use.[18] Failures of oversight can lead to operational instability, regulatory action, and loss of trust. In the age of AI, fiduciary duty means owning the ethical framework that governs its use.

This convergence of law, ethics, and governance broadens fiduciary responsibility. Acting with care and diligence now extends beyond financial statements and market assessments to include scrutiny of algorithmic outputs, training data, and automated decisions. Acting in good faith means balancing profit with customer trust and regulatory expectations. Avoiding misuse of position now includes not exploiting privileged access to data or technology for personal benefit.

The law does not require perfection. Directors are not liable for every misstep, nor are they guarantors of technological outcomes. But the law does require engagement, diligence, and active oversight. It requires boards to be inquisitive rather than passive, to be informed rather than indifferent. It requires directors to acknowledge that AI is not an optional add-on to governance, but a central feature of modern corporate responsibility.

The standard is evolving, but the underlying principle is timeless. Fiduciary duty is about stewardship. Directors are stewards of capital, of reputation, of trust. In the age of AI, they are also stewards of data, algorithms, and digital systems that shape the lives of customers and the fabric of society. The responsibility is weighty, but it is not optional. Boards that rise to this challenge will not only protect themselves from liability but also build stronger, more resilient organizations.

Reference

1. Lexology, 'Fiduciary Duties in the Age of Artificial Intelligence' (2024) https://www.lexology.com/library/detail.aspx?g=e70b0ed6-176e-4b34-aec6-12f594006dc8.
2. *Corporations Act* 2001 (Cth).
3. John Micklethwait and Adrian Wooldridge, The Company: A Short History of a Revolutionary Idea (Modern Library, 2005).
4. In re Caremark International Inc Derivative Litigation 698 A.2d 959 (Del. Ch. 1996).
5. Jarrod Anderson, *The Chief AI Officer's Handbook* (Packt Publishing, 2025).
6. Corporations Act 2001 (Cth).
7. Phillip Lipton et al, *Understanding Company Law* (Thomson Reuters, 19th ed, 2018).
8. Ibid.
9. *Corporations Act* 2001 (Cth).
10. Lexology, 'Fiduciary Duties in the Age of Artificial Intelligence' (2024) https://www.lexology.com/library/detail.aspx?g=e70b0ed6-176e-4b34-aec6-12f594006dc8.
11. *ASIC v RI Advice Group Pty Ltd* [2022] FCA 496.
12. *ASIC v Healey* [2011] FCA 717.
13. Ibid.
14. João da Silva, FTX says it has billions more than owed to victims(BBC, 2024) https://www.bbc.com/news/articles/c9wz3wkrdj90
15. Daniel Jurafsky and James H Martin, *Speech and Language Processing* (3rd ed, draft, Prentice Hall, 2023).
16. Jarrod Anderson, *The Chief AI Officer's Handbook* (Packt Publishing, 2025).
17. ISO/IEC 42001:2023 – Artificial Intelligence Management System Standard.
18. Department of Industry, Science and Resources (Cth), Voluntary AI Safety Standard (Australian Government, 2024) https://www.industry.gov.au/publications/voluntary-ai-safety-standard.

"Governance is the Difference between Steering technology and being steered by It."

CHAPTER 12

Governance

Governance defines how an organization is directed, controlled, and is held accountable.[1] At its best, it provides clarity about who makes decisions, how they are made, and how outcomes are monitored. Good governance is more than compliance. It ensures that decisions align with the organization's purpose, strategy, and risk appetite while building trust, transparency, and fairness.

In the age of artificial intelligence, where there is limited legislative support, governance is more important than ever.[2] Technology has amplified the speed, scale, and consequences of decision-making. An AI model that produces biased results, a cloud platform that is misconfigured, or a dataset that is mishandled can create risks which has a significant impact. This means that governance must extend into the systems, data, and digital processes that shape how organizations operate.

This chapter approaches governance through five lenses: corporate, enterprise, IT, data, and AI. Each are connected by the same core principles of transparency, accountability, fairness, and responsibility. Corporate governance provides the foundation, clarifying the relationship between boards, management, and stakeholders. Enterprise governance adds the dimension of balancing performance and conformance, ensuring that growth and compliance move together. IT governance addresses the systems and technology decisions that enable strategy but also introduce risk. Data governance manages the lifeblood of modern organizations, making sure information is accurate, secure, and ethically used. Finally, AI governance builds on all of these traditions but adapts them to the unique risks and uncertainties of intelligent systems.

Board members are not expected to become experts in every domain. Their role is to ensure that governance principles are applied consistently, whether the decision involves approving a balance sheet, investing in digital infrastructure, or deploying AI into customer-facing operations. Governance is the framework that connects oversight with accountability, risk with opportunity, and technology with trust. It is the compass that allows directors to navigate complexity without losing sight of the organization's purpose.

Types of Governance

Corporate Governance

Corporate governance provides the overarching framework for how an organization is directed, controlled, and held accountable.[3] It sits at the intersection of law, strategy, and ethics, clarifying the relationship between the board, management, shareholders, and stakeholders. At its heart, it defines who has authority, how that authority is checked, and how responsibility is enforced.

The roots of corporate governance lie in agency theory, which explains the gap between shareholder interests and managerial goals.[4] Governance exists to close that gap by placing duties of care, loyalty, and diligence on directors. Over time, thinking expanded. Stewardship theory reframe directors as custodians of long-term health, not just monitors of management. Stakeholder theory went further, recognizing that companies impact employees, customers, regulators, and communities.[5] This is even more relevant today as AI systems affect not only shareholders but also wider society.

Corporate governance principles are embedded in global and domestic frameworks such as the OECD Principles of Corporate Governance and, in Australia, the ASX Corporate Governance Council's Principles of Good Corporate Governance and Best Practice Recommendations first published in 2003.[6] These frameworks continue to shape how boards exercise authority and accountability in modern contexts. Four principles stand out across both: transparency, accountability, fairness, and responsibility.[7] These principles guide directors to oversee risk, approve and monitor strategy, appoint and hold executives accountable, and safeguard financial integrity through transparent reporting and independent audit.

What makes these principles fundamental is their adaptability. Transparency now extends beyond financial disclosure to include algorithmic explainability, traceability of data, and visibility of third-party AI dependencies. Accountability no longer stops at management but extends to how decisions are delegated to systems or vendors. Fairness includes ensuring that automated outcomes do not embed bias or discrimination. Responsibility demands that directors understand enough about the technologies they authorize to exercise informed judgment.[8] Each principle remains the same in spirit but must now operate in a landscape defined by data flows, algorithms, and machine learning. In this environment, ethical governance becomes the practical expression of these long-standing ideals.

The digital era has expanded governance beyond financial reporting and executive oversight into data governance, cybersecurity, digital transformation, and the ethics of artificial intelligence. A board that fails to interrogate the risks of deploying an untested AI model into customer-facing systems may be as negligent as one that fails to scrutinize a balance sheet. Similarly, ignoring cybersecurity threats or privacy obligations can represent a breach of duty of care under modern interpretations of directors' responsibilities.

History shows the consequences of failure. The collapse of Enron in 2001 exposed the cost of opacity and weak oversight,[9] while the implosion of FTX in 2022 showed that even digital-era firms cannot abandon transparency,[10] independent oversight, or robust risk management without catastrophic consequences.

In the age of AI and digital transformation, these lessons carry forward. Fiduciary and governance duties cannot be delegated to algorithms, data scientists, or vendors. Boards remain ultimately responsible for ensuring that technologies are deployed responsibly, that risks are managed proactively, and that ethical and legal obligations are upheld.

Corporate governance is not static nor is it historical. Its fundamentals of transparency, accountability, fairness, and responsibility remains constant, but boards must now apply these principles to new frontiers such as digital assets and AI-driven decision-making. These failures can erode trust at unprecedented speed. Successful Corporate Governance is grounded in resilience, trust, and the ability to navigate complexity while staying true to the organization's purpose.

Enterprise Governance

If corporate governance is the foundation, enterprise governance is the structure built upon it. It balances performance, conformance, and responsibility across the organization. [11] Businesses cannot rely on compliance alone to survive, nor on performance alone to succeed. To maintain trust, boards must align purpose, strategy, and societal expectations within a single, integrated system of governance.

Enterprise governance defines direction, allocates resources, oversees risk, and ensures leaders are accountable. While corporate governance focuses on compliance and accountability, enterprise governance extends to long-term value creation.[12] It asks whether growth is sustainable, whether risks are being managed responsibly, and whether the organization is meeting its obligations to shareholders, employees, regulators, and society. [13]

Enterprise governance unites two dimensions: conformance and performance. Conformance ensures compliance and accountability. Performance drives innovation, growth, and value over time. Both are essential. Compliance without innovation creates stagnation. Performance without oversight erodes trust.

The failures at Rio Tinto's Juukan Gorge in 2020 illustrate what happens when conformance is satisfied but responsibility and accountability are ignored.[14] The destruction of sacred Aboriginal caves was legally permissible under heritage approvals, but it failed the test of responsibility and undermined the company's social license to operate. The result was global outrage, the resignation of senior executives, and long-lasting reputational damage.[15]

Enterprise governance is not about doing the minimum required by law but about aligning decisions with societal expectations, cultural sensitivity, and long-term sustainability. It requires directors to integrate strategic direction, value creation, and risk oversight[16], applying these disciplines not only to traditional areas but also to emerging domains such as cybersecurity, AI adoption, and climate risk.

Enterprise governance therefore represents more than a conceptual model. It is a practical guide to decision-making.[17] It forces directors to confront trade-offs between compliance and performance, between short-term gains and long-term sustainability, between shareholder value and stakeholder trust. It requires the courage to say no to strategies that may be profitable in the near term but damaging in the long run, and the foresight to invest in initiatives that strengthen resilience even if they take time to deliver returns.

The link to technology and AI cannot be avoided. Enterprise governance provides the lens through which emerging technologies must be assessed.[18] Performance demands that organizations leverage AI for efficiency, innovation, and competitive advantage. Conformance demands that they comply with laws, regulations, and internal policies, from privacy obligations to discrimination laws. Responsibility demands that they adopt AI in a way that is fair, transparent, and accountable to the communities they serve.[19] Enterprise governance is thus the bridge between risk and strategy. It ensures that AI adoption is not an isolated IT project but part of the organization's holistic approach to value creation and accountability.

Enterprise governance is the architecture that allows organizations to operate effectively, ethically, and sustainably in a world where risks and responsibilities are increasingly complex. It integrates corporate oversight with strategic management, balancing conformance and performance in service of long-term value. Embracing enterprise governance is not optional. It is a prerequisite for credibility, resilience, and trust.

IT Governance

IT governance is the framework that defines decision-making rights and accountability for the use of IT resources.[20] Its purpose is to align IT strategy with business objectives, ensure value delivery, and manage risk. In many organizations, IT represents one of the largest capital expenditures, and its role has shifted from a support service to a strategic enabler. The systems that once ran payroll and email now host customer platforms, sensitive health records, and AI models that shape real-time business decisions.

This shift has elevated IT governance to a critical boardroom discussion. Frameworks such as COBIT and ISO/IEC 38500 provide comprehensive standards that clarify roles and responsibilities for boards and executive leadership.[21] These frameworks are not intended to turn directors into technologists but to empower them to ask the right questions necessary for robust oversight.

For example, boards are guided to inquire who holds decision-making authority over IT investments, how these investments align with overall business goals, by what criteria value is measured, and how associated risks are identified and managed.

This structure is essential for establishing clear ownership and accountability. It is important for boards to understand whether cybersecurity risks fall under the Chief Information Officer (CIO), the

Chief Information Security Officer (CISO), or between both roles. Clarity is also needed on who holds ultimate authority when AI systems handle sensitive customer data.

The accountability can be role-specific and significant. For example, in 2023, the U.S. Securities and Exchange Commission (SEC) charged SolarWinds' CISO, Timothy Brown, with fraud and internal control failures relating to a 2020 cyber-attack.[22]

This case highlights that accountability does not only rest with the company as an entity but can be specifically attributed to individuals in key cybersecurity leadership positions. Clear governance frameworks that delineate roles such as CIO and CISO are therefore vital to ensure responsibility is properly assigned and managed.

The significance of IT governance has increased dramatically with digital transformation. As organizations migrate to the cloud, adopt agile development, and deploy AI systems, risks multiply: cyber threats, misconfigurations, privacy violations, and vendor dependencies. Boards that once regarded IT as a back-office concern must now treat it as a board-level issue.

The 2022 Optus breach[23] illustrate how failures in IT governance can escalate into national crises. In each case, the problem was not only technical but also governance because decision rights and oversight were unclear.

IT governance also extends to resilience. Cyber incidents, ransomware attacks, and outages all test whether governance structures are robust. A well-designed governance model ensures that contingency planning, backup strategies, and vendor contracts are not buried in the IT department but reported transparently to the board. Boards that fail to oversee IT governance risk not only financial loss but also regulatory penalties and reputational harm. Regulators globally, from APRA in Australia to the SEC in the United States, are making it explicit that directors are accountable for digital oversight.

Ultimately, IT governance extends corporate governance. Its purpose is not technology for its own sake, but ensuring technology serves the long-term interests of the organization, its customers, and its stakeholders. In the digital age, IT governance is no longer optional. It is a core fiduciary duty.

Data Governance

While IT governance focuses on the broader framework that focuses on decision-making rights and responsibilities linked to an organization's IT resources and systems, data governance is more specialized. Fundamentally, data governance Data governance involves setting policies and procedures to maintain data quality, protect privacy, manage metadata, and ensure regulatory compliance with respect to data handling. It defines roles such as data owners and stewards responsible for data accuracy and security.

The urgency has grown as the volume and variety of data expand at unprecedented speed.[24] Every transaction, email, sensor reading, and social interaction generates new records, and AI accelerates this further by consuming vast datasets and producing synthetic data. Scale and diversity make governance essential.

Data now spans personal information, financial records, intellectual property, health data, and biometric identifiers.25 Each carries unique ethical and legal considerations. Questions of how personal is too personal, how sensitive is too sensitive, and when data should be deleted rather than stored are not for data scientists alone. They are governance questions bound by law and expectation.

The ethical dimension of data governance is central to responsible data use. AI systems are only as fair as the data they are trained on. Poorly governed data can entrench bias, exclude vulnerable groups, or misuse personal information in ways that damage public trust and organizational reputation. Regulations such as the EU's GDPR and Australia's Privacy Act reforms codify some obligations, but ethical responsibility goes beyond compliance.[26] Governance operationalizes ethics by ensuring that data is collected with consent, processed transparently, and used consistently with corporate values and societal norms.

Good data governance is also crucial for unlocking value and driving innovation. When data is trusted, well-catalogued, and accessible under clear rules, it empowers staff and partners to use data responsibly and creatively without fear of breaching compliance.[27] Strong governance provides clarity and confidence. The alternative is data hoarding, shadow datasets, and lost opportunity because data remains inaccessible, untrusted, or unusable.

Every AI system depends on the data it is trained on. This includes both the inputs it receives and the outputs it generates. Weak governance leads directly to weak outcomes. The result is bias embedded in training sets, hallucinations amplified by poor inputs, and sensitive information exposed through unmonitored prompts.

Strong governance turns AI into a strategic advantage. Well-classified, ethically sourced, and properly stewarded data produces models that are accurate, compliant, and trusted.[28] Investing in governance enables AI that is explainable, trustworthy, and aligned with the organization's purpose.

AI Governance

AI governance and data governance are closely related but serve distinct purposes within an organization's governance framework. Data governance manages the organization's information assets, ensuring their quality, security, privacy, and proper lifecycle management. It defines policies and assigns roles for data collection, storage, access, and compliance so that data remains accurate, trustworthy, and handled responsibly.

In contrast, AI governance is emerging as a key framework for modern boards.[29] It is narrower in scope than corporate or enterprise governance but far more complex in practice. Unlike traditional IT systems, AI is dynamic, adaptive, and often lacks transparency. Models can drift, datasets can shift, and outputs can change without notice.

This complexity requires governance frameworks that build on established principles while addressing AI's unique ethical, operational, and risk challenges. Successful AI governance requires boards to treat oversight of AI with the same rigor they apply to financial controls and strategic risk management.

The foundations of AI governance build on principles already well established in other frameworks. Corporate governance sets expectations of good faith, accountability, and acting in the company's best interests.[30] Enterprise governance reminds boards that growth and profit matter, but not if they come at the cost of breaking rules or losing public trust. IT governance shapes decision-making, risk oversight, and strategic alignment for technology investments. Data governance sets the standards for accuracy, security, and ethical data use.[31] Together, these create a unified discipline that AI governance extends into new territory.

AI introduces a fundamental difference. It learns and evolves. It produces outcomes that can shift over time and sometimes extend beyond human visibility or control. That unpredictability demands clear boundaries. Policies must define what AI can do, where it may be used, and what risks are unacceptable.

Organizations often deploy AI opportunistically, driven by competitive pressure or fear of missing out, without defining purpose or boundaries. Boards should insist on formal AI policies that set out the limits of use, permissible applications, and unacceptable practices. For example, a policy might allow AI in customer service but prohibit it in employee monitoring or high-stakes financial decisions without additional oversight.[32] Scope must extend across in-house systems and third-party tools, and objectives must tie directly to business strategy. If AI does not advance purpose, create value, or strengthen resilience, it risks becoming a liability rather than a driver of performance.

Artificial intelligence should not be treated as an experiment or pursued simply to cut costs. AI governance must be integrated with business strategy and plans. Boards should ensure that AI initiatives align with the organization's mission, vision, and long-term objectives. AI investments should be part of capital planning, resource decisions, and performance targets, just like any other major strategic initiative.

A strong governance framework ensures that projects are chosen for their contribution to long-term value rather than short-term hype. Operating plans should show how AI adoption will be managed, including staff training, workflow adjustments, and continuous monitoring of outputs and emerging risks. Boards do not need to manage operational details themselves, but they should require evidence that AI is being embedded responsibly and strategically across the organization.

Roles and responsibilities are another pillar of AI governance. In many organizations, AI is still treated as a technical experiment left to data scientists or IT departments. This is a mistake. Without clear ownership, accountability becomes obscured, and when things go wrong, blame is easily shifted. AI governance must assign responsibility across all levels.

Boards provide oversight, executives execute strategy, and cross-functional committees manage the operational and ethical complexities. Project management frameworks such as RACI (Responsible, Accountable, Consulted, Informed) can be adapted to clarify roles and responsibilities.[34] For example, the board may be accountable for approving overall AI risk appetite, executives may be responsible for implementing policies, compliance and risk teams may be consulted on controls, and staff across functions must be informed about appropriate use. What matters is that accountability is never lost in technical complexity. If an AI system produces harm, there must be clarity about who answers for it.

Defining risk appetite is critical. Just as boards articulate their tolerance for financial, reputational, or operational risk, they must also communicate their appetite for AI risk.[35] This includes setting thresholds for acceptable error rates, tolerance for bias in outcomes, willingness to accept opaque or "black box" systems, and expectations for transparency with customers. Risk assessments are equally important. They should be ongoing evaluations that test for bias, check explainability, monitor resilience, and ensure compliance with evolving regulations.

Transparency remains one of the most important safeguards in AI oversight. Without it, organizations may not fully understand how third-party systems operate or what data they rely on. Vendors might use training data that clients cannot review, introducing hidden bias or security weaknesses. In some cases, AI features are enabled without clear disclosure, leaving companies unaware of the risks until an issue arises.

The 2025 Salesforce incident, where attackers exploited OAuth tokens from a third-party app to exfiltrate large volumes of customer data, highlighted the need for stronger vendor visibility.[36] In another example, McDonald's launched a chatbot that inadvertently exposed employee credentials.[37] These cases underline why boards must have clear sight of all AI systems in use, whether developed internally or supplied by vendors.

An AI inventory is essential for good governance. It should list every AI tool the organization uses, whether built in-house, bought from a vendor, or still in testing.[38] For each tool, record what it is used for, what data it relies on, the risks it carries, and the safeguards in place. For example, an inventory might record that a chatbot is powered by a third-party language model, trained on proprietary datasets, exposed to customers, and monitored for bias.

Lifecycle monitoring is where AI governance most clearly differs from traditional IT oversight. Traditional systems are largely static, with controls checked at deployment and updated periodically. AI, by contrast, continues to evolve after release. As models learn and adapt, their behavior can shift, and performance may deteriorate over time. Governance must therefore include continuous monitoring from development through deployment, retraining, and eventual retirement.[39] Oversight should track performance, fairness, security, and compliance, with regular reporting to the board. Escalation must be mandatory. When bias or unexplained change is detected, responsibility should move quickly through defined channels to ensure the board is informed and corrective action is taken.

AI governance is not just a technical tool. It affects people. Within the workplace, it can reshape roles, workflows, and the skills employees need. It also influences the experiences of customers, determining outcomes such as credit approvals, job opportunities, insurance assessments, and access to services. At a broader level, its impact extends to society, shaping trust, fairness, and even democratic integrity. Governance must therefore promote fairness in the workforce, transparency in customer outcomes, and accountability for wider social consequences.

Few organizations develop AI entirely in-house. Most depend on external vendors for APIs, pre-trained models, or off-the-shelf software.[40] These products often include embedded AI features, sometimes activated without the organization's knowledge. This creates an "AI supply chain" that may not be fully visible to boards or management. The risk is that vendors could use undisclosed datasets, introduce bias, or manage data in ways the organization did not anticipate. Boards must ensure that due diligence extends to AI by requiring transparency from vendors, requesting explanations of how AI is used, and including appropriate protections in contracts.

Several frameworks have emerged to help executives structure AI governance. Deloitte's AI Governance Roadmap, the NIST AI Risk Management Framework, the EU AI Act, and the forthcoming ISO/IEC 42001 all provide useful structures for aligning strategy, managing risk, and embedding accountability.[41] These frameworks are valuable references, but boards should adapt them to their own context rather than adopt them wholesale. Not all businesses are the same, and maturity levels vary. Boards should endorse them as guides while recognizing they may never be a perfect fit.

Traditional governance cannot keep up with the pace of AI. Models can drift within weeks, regulations change almost as quickly, and new risks like deepfakes or data poisoning emerge faster than most organizations can respond. AI governance must be dynamic. Boards should expect frequent reviews, ongoing monitoring of high-risk models, and oversight that adapts as technology shifts. Governance that stands still becomes the very risk it was designed to prevent.

PUTTING GOVERNANCE TO WORK

Effective governance takes practice. It lives in policies, committees, reporting lines, and the culture set by directors and executives.[42] The "tone from the top" shapes how accountability is understood and applied, creating the conditions in which it either flourishes or fails.

Strong governance is often misunderstood as bureaucracy, when in fact it enables action. Clear rules and well-defined processes give staff confidence, reassure investors, and reduce the likelihood of regulatory intervention. Good governance does not slow an organization down; it lowers friction by replacing uncertainty with clarity. It gives leaders a structure through which decisions can be made quickly, transparently, and with shared understanding.[43]

AI, however, challenges the assumptions that traditional governance was built on. It operates at scale, makes probabilistic decisions, and often learns in ways that are difficult to predict or explain. These characteristics demand stronger oversight, new reporting structures, and greater technical fluency at the board level. Boards cannot rely solely on existing corporate or IT governance frameworks to manage AI risk. They must extend those structures to cover emerging issues such as bias, drift, and opaque decision-making, while ensuring that accountability remains visible and traceable.

This is where governance becomes practical. Policies must be translated into action through committees, reporting frameworks, and vendor oversight. The sections that follow explore how boards can adapt their structures and expectations to govern AI responsibly and to ensure that technology supports, rather than undermines, trust in leadership.

Third party

I have already touched on third-party risk throughout this book, but it deserves closer attention. The relationship between an organization and its AI vendors is no longer transactional. It is strategic, often carrying the same weight as internal infrastructure or core systems. Once AI is embedded through a third-party platform, it becomes part of the organization's operational fabric, shaping workflows, decision-making, and even culture. That level of dependency demands strong governance, not blind trust.

When organizations partner with vendors such as Microsoft, Salesforce, or Google, they also inherit the risks that come with them.[42] Vendors may switch on new AI features without warning, change how data is processed, or update models in ways that ripple through connected systems.[44] These shifts can create sudden exposures, particularly when personal or sensitive data is involved. Even when the vendor fails to notify clients, accountability for misuse or breach still sits with the organization that chose to adopt the service. Regulators and customers rarely distinguish between who built the system and who used it.

Contracts remain essential but are no longer sufficient on their own. [45] The speed of AI innovation has outpaced the language in most legacy agreements. Few contracts address how models are retrained, what data can be used in continuous learning, or who is accountable for harm caused by model error. Without clear terms, liability can quietly shift back onto the business.

Contracts should be updated to reflect AI-specific risks. Agreements must set out obligations for security, transparency, and timely notification of incidents or significant model changes. They should also clarify how liability is shared or transferred when personal or regulated data is involved.[46] The board should take comfort that vendors hold adequate insurance for AI-related breaches and whether any gaps could leave the organization exposed.

Additionally, vendors should be evaluated against established standards such as ISO 27001, the NIST Cybersecurity Framework, or industry-specific benchmarks.[47] Independent audits, penetration tests, and risk assessments should be ongoing, not occasional. These exercises provide assurance that vendor claims about privacy and security are supported by evidence rather than marketing.
Furthermore, organization should be assured that vendors monitor and control their own AI systems. This includes understanding how training data is sourced, how bias and drift are detected, and what governance applies to model updates. Transparency in these areas is essential for assessing both operational and ethical risk.

Adopting third-party AI is not only a technology decision. It is a governance one. Contracts must evolve as fast as the tools they govern. Due diligence must be continuous. Oversight must extend beyond vendor assurances. Without these measures, outsourcing does not remove risk; it merely relocates it, making it difficult to track and manage.

When Third-Party AI Becomes a Liability

In recent months, several high-profile failures have shown how fragile third-party AI systems can be and how quickly those weaknesses spill over into the organizations that rely on them. Salesforce's Agentforce, for example, was found to contain a critical flaw now known as "ForcedLeak."[48] Attackers were able to plant malicious instructions into an online lead form which the AI agent later processed. Without any additional hacking, the system obediently retrieved sensitive CRM data and sent it to an attacker's server. Salesforce responded with patches and new controls, but the episode revealed how easily an embedded AI tool can be manipulated once it is tied into core business processes.

A second case involved Microsoft's GitHub Copilot. Security researchers discovered that code repositories once made public and later marked private were still being accessed through Copilot because cached and indexed versions had not been purged.[49]

Developers assumed their data was secure, but the AI tool continued to draw from a shadow copy. This revealed a deeper problem. Once information enters an AI system, control over that data can become uncertain. Permission changes may not take effect immediately, and the line between public and private information is often thinner than many organizations realize.

The third incident came from Microsoft's own Copilot Enterprise service.[50] A vulnerability was uncovered that allowed attackers to escalate privileges and gain root access to the backend container environment. If exploited, such access would have handed control of internal processes and data to the attacker, bypassing normal security layers altogether. It highlighted that even the most reputable vendors can become a direct source of compromise if their AI services are not rigorously secured.

Taken together, these incidents highlight an important reality. Relying on third-party AI does not eliminate risk. It simply changes where that risk sits. Vendors may control the code and infrastructure, but customers remain accountable when failures occur.[51]

AI services that are embedded through contracts must be reviewed with the same rigor as in-house systems. Oversight cannot end with licensing terms or polished vendor briefings. It needs to include an understanding of how vendors monitor their tools, how quickly they patch vulnerabilities, and whether permission changes are properly enforced across all layers of their platforms. Without this discipline, organizations inherit their partners' weaknesses while carrying the full weight of responsibility when things go wrong.

From my perspective, the cases of Salesforce, GitHub Copilot, and Microsoft show how quickly vendor AI risks can appear and how disruptive they can become. Each incident is a reminder that contracts and frameworks are important but never enough.[52] Oversight must be built into board structures with clear lines of reporting and accountability. The real question is not whether oversight exists but where it sits and whether those responsible understand the risks they are meant to govern.

How Boards should Steer AI

Effective AI governance cannot live in policy documents alone. Policies and principles written in glossy reports mean little unless they are anchored in clear oversight structures.[53] The central question is not only what governance exists but where it sits. Without defined accountability, AI risk becomes everyone's problem and therefore no one's responsibility. Boards already understand this challenge in other areas. Financial oversight sits with the audit committee. Cybersecurity is often managed by the risk committee.[54] People and remuneration practices are handled by their own dedicated group.

Each of these domains has a home where issues are surfaced, debated, and escalated. AI deserves the same clarity. When it shifts from one department to another, treated one month as technology, the next as compliance, and later as innovation, directors never see a complete picture of risk or opportunity.

For organizations at an early stage of AI adoption, folding oversight into the existing risk or audit committee may be enough. These committees already examine systemic risks and controls, and AI can be considered alongside cyber, privacy, and operational matters. This approach helps prevent fragmentation and ensures AI remains part of the wider enterprise risk portfolio.

As adoption matures, a dedicated committee may be need. A technology or ethics committee signals to staff, regulators, and investors that AI is not just another IT project but a strategic priority with its own governance lens. Committees of this nature can explore topics such as algorithmic fairness, explainability, model drift, and ethical use in greater depth than a general risk committee might allow. The right structure depends on the organization's scale, its exposure, and the expectations of regulators.

Regardless of the model chosen, directors must also insist on practical tools that translate technical performance into clear oversight.[55] Boards often receive AI updates that are either highly technical or so general that they add no real value. Discussions about parameters or training sets rarely help, and vague assurances that "the system is working well" are equally unhelpful. What boards need are dashboards that convert complex indicators into language that supports sound business judgment.

A well-designed AI dashboard should highlight drift indicators showing when accuracy is declining, fairness metrics revealing differences across user groups, and incident summaries detailing near misses or harmful outputs.[56] It should also map where third-party models are embedded into key business processes. These insights give directors a grounded view of how AI systems are performing and where risks may be forming.

The value of reporting is not in the numbers themselves but in how they guide action. Dashboards give boards the visibility to ask informed questions, allocate resources, and hold management accountable. When oversight becomes continuous, not episodic, governance moves from paper to practice. That is where effective AI leadership begins.

Reference

1. Organisation for Economic Co-operation and Development (OECD), G20/OECD Principles of Corporate Governance (2023).
2. Jarrod Anderson, The Chief AI Officer's Handbook (Packt Publishing, 2025).
3. R Iskander and N Chamlou (eds), Corporate Governance: A Framework for Implementation (World Bank, 2000) Foreword by Sir Adrian Cadbury.
4. Michael C Jensen and William H Meckling, 'Theory of the Firm: Managerial Behavior, Agency Costs and Ownership Structure' (1976) 3(4) Journal of Financial Economics 305.
5. Lex Donaldson and James H Davis, 'Stewardship Theory or Agency Theory: CEO Governance and Shareholder Returns' (1991) 16(1) Australian Journal of Management 49.
6. Organisation for Economic Co-operation and Development (OECD), G20/OECD Principles of Corporate Governance (2023).
7. Robert P Austin and Ian M Ramsay, Principles of Corporations Law (Lexis Nexis, 2018)
8. Ibid.
9. Federal Bureau of Investigation (FBI), Enron (History, FBI) https://www.fbi.gov/history/famous-cases/enron.
10. Douglas W Arner et al, 'The Financialization of Crypto: Lessons from FTX and the Crypto Winter of 2022–2023' [2023] UNSW Law Research Series 31.
11. Steven De Haes, Wim Van Grembergen, Anant Joshi and Tansu Çalişir, Enterprise Governance of Information Technology: Achieving Alignment and Business Value (Springer, 2020).
12. Ibid.
13. Robert Tricker, Corporate Governance: Principles, Policies and Practices (5th ed, Oxford University Press, 2023).
14. Australians for Native Title and Reconciliation (ANTaR), 'The Destruction of Juukan Gorge' (2024) https://antar.org.au/issues/cultural-heritage/the-destruction-of-juukan-gorge/.
15. Ben Butler et al, 'Rio Tinto CEO and Senior Executives Resign from Company after Juukan Gorge Debacle' (The Guardian, 2020) https://www.theguardian.com/business/2020/sep/11/rio-tinto-ceo-senior-executives-resign-juukan-gorge-debacle-caves.

16. Steven De Haes, Wim Van Grembergen, Anant Joshi and Tansu Çalişir, Enterprise Governance of Information Technology: Achieving Alignment and Business Value (Springer, 2020).
17. Ibid.
18. Jarrod Anderson, The Chief AI Officer's Handbook (Packt Publishing, 2025); ISACA, COBIT 2019 Framework: Governance and Management Objectives (2019).
19. Brenda Van Rensburg, Submission to the Department of Industry, Science and Resources — Supporting Responsible AI Consultation (Australian Government, 2024) https://consult.industry.gov.au/supporting-responsible-ai/submission/view/63.
20. Yassine Maleh et al, IT Governance and Information Security: Guides, Standards, and Frameworks (CRC Press, 2022).
21. IT Governance, What is ISO? (website viewed August 2025) IEC 38500 https://www.itgovernance.co.uk/iso38500
22. Betsy Atkins, SEC Charges CISO With Fraud In Landmark Cybersecurity Case (2023) https://www.forbes.com/sites/betsyatkins/2023/11/16/sec-charges-ciso-with-fraud-in-landmark-cybersecurity-case/
23. Slater Gorden, Optus Data Breach Class Action (2023) https://www.slatergordon.com.au/class-actions/current-class-actions/optus-data-breach
24. Jarrod Anderson, The Chief AI Officer's Handbook (Packt Publishing, 2025).
25. Optus, 'Optus Notifies Customers of Cyberattack Compromising Customer Information' (2022) https://www.optus.com.au/about/media-centre/media-releases/2022/09/optus-notifies-customers-of-cyberattack.
26. Anita Kibunguchy et al, Data Governance: The Definitive Guide (O'Reilly Media, 2023).
27. Viktor Mayer-Schönberger and Kenneth Cukier, Big Data: A Revolution That Will Transform How We Live, Work, and Think (John Murray, 2013).
28. National Institute of Standards and Technology (NIST), Artificial Intelligence Risk Management Framework (AI RMF 1.0) (NIST Special Publication No 1270, January 2023) https://nvlpubs.nist.gov/nistpubs/ai/NIST.AI.100-1.pdf
29. Ibid.

30. Robert P Austin and Iam Ramsay, Principles of Corporats Law (2023, Lexis Nexis Butterworths)
31. Anita Kibunguchy et al, Data Governance: The Definitive Guide (O'Reilly Media, 2023).
32. Emily M Bender and Alex Hanna, AI Con: Who Owns the Future of Artificial Intelligence? (W W Norton & Company, 2024).
33. Project Management Institute (PMI), A Guide to the Project Management Body of Knowledge (PMBOK® Guide) (7th ed, Project Management Institute, 2021).
34. Paul Daugherty and H James Wilson, Radically Human: How New Technology Is Transforming Business and Shaping Our Future (Harvard Business Review Press, 2022).
35. Ravie Lakshmanan, 'Salesforce Patches Critical ForcedLeak Bug Exposing CRM Data via AI Prompt Injection' (The Hacker News, 2025) https://thehackernews.com/2025/09/salesforce-patches-critical-forcedleak.html.
36. Tony Bradley, McDonald's AI Breach Reveals The Dark Side Of Automated Recruitment (2025) https://www.forbes.com/sites/tonybradley/2025/07/15/mcdonalds-ai-breach-reveals-the-dark-side-of-automated-recruitment/
37. Jarrod Anderson, The Chief AI Officer's Handbook (Packt Publishing, 2025).
38. International Organization for Standardization, ISO/IEC 42001:2023 – Artificial Intelligence Management System Standard (ISO, 2023).
39. Glenn M Dotson, Practical Cloud Security (2nd ed, O'Reilly Media, 2024).
40. Australian Institute of Company Directors (AICD), Good Governance Principles and Guidance for Not-for-Profit Organisations (2021) or Guiding Principles of Good Governance
41. Ibid.
42. Jarrod Anderson, The Chief AI Officer's Handbook (Packt Publishing, 2025).
43. Debevoise, Managing AI-Related Risks Associated with Vendors (2024) https://www.debevoise.com/insights/publications/2024/12/managing-ai-related-risks-associated-with-vendors
44. KPMG, "Securing and Understanding the Risk of Third-Party AI Use" (2025) https://kpmg.com/us/en/articles/2025/securing-understanding-risk-third-party-ai-use.html
45. Dilan Thampapillai and Alex Bruce, Contract Law: Text and Cases (3rd ed, Lexis Nexis, 2021).

46. Khatri, V. & Brown, C. V. (2010). Designing Data Governance. Communications of the ACM, 53(1), 148–152.
47. National Institute of Standards and Technology (NIST), Artificial Intelligence Risk Management Framework (AI RMF 1.0) (US Department of Commerce, 2023).
48. Ravie Lakshmanan, 'Salesforce Patches Critical ForcedLeak Bug Exposing CRM Data via AI Prompt Injection' (The Hacker News, 2025) https://thehackernews.com/2025/09/salesforce-patches-critical-forcedleak.html.
49. Ziv Karliner, 'New Vulnerability in GitHub Copilot and Cursor: How Hackers Can Weaponize Code Agents' (Pillar Security Blog, 2025) https://www.pillar.security/blog/new-vulnerability-in-github-copilot-and-cursor-how-hackers-can-weaponize-code-agents.
50. Erik van Klinken, 'Critical Vulnerability Discovered in Copilot Enterprise' (Techzine, 2025) https://www.techzine.eu/news/security/133337/critical-vulnerability-discovered-in-copilot-enterprise/.
51. Jarrod Anderson, The Chief AI Officer's Handbook (Packt Publishing, 2025).
52. World Economic Forum. (2022). Empowering AI Leadership: Board Governance of AI [Report]. Retrieved from https://www3.weforum.org/docs/WEF_Empowering_AI_Leadership_2022.pdf
53. Organisation for Economic Co-operation and Development (OECD), G20/OECD Principles of Corporate Governance (2023).
54. World Economic Forum. (2022). Empowering AI Leadership: Board Governance of AI [Report]. Retrieved from https://www3.weforum.org/docs/WEF_Empowering_AI_Leadership_2022.pdf
55. World Economic Forum. (2022). Empowering AI Leadership: Board Governance of AI [Report]. Retrieved from https://www3.weforum.org/docs/WEF_Empowering_AI_Leadership_2022.pdf
56. ASX Corporate Governance Council, Principles of Good Corporate Governance and Best Practice Recommendations, (2019) https://www.governanceinstitute.com.au/advocacy/insights-asx-corporate-governance-principles-and-recommendations-fourth-edition-2019/

"The true test of ethical accountability is whether you would defend the Decision in public as proudly as you made it in Private."

CHAPTER 13

Ethical Accountability

Ethics sits in the space between law and governance.[1] Where law asks, "what is permitted?" and governance asks, "how do we oversee it?", ethics asks a deeper question: "what is right?". It is not about compliance with a statute, but about the expectation that organizations will act with integrity even when the law is silent or slow to respond.

In Australia, the Constitution (s 51) empowers the Commonwealth to make laws for "peace, order, and good government."[2] But it was never intended to micromanage every aspect of corporate behavior. Legislation provides the baseline for what must be done, not the boundary of what can be achieved. Flowing from this, Governance frameworks provide the structures for accountability. Ethics fills the middle ground, ensuring decisions are made with foresight, fairness, and legitimacy.

Boards often encounter situations where compliance is met but the outcome raises ethical concerns. The Cambridge Analytica case is a reminder of poor ethical decisions. Most of the company's actions were technically lawful, yet a few poor decisions led to a global scandal.[3] What began as an app, 'This Is Your Digital Life', quietly harvested user data that was later used to influence the 2016 U.S. presidential campaign.[4] Was the company collecting data for a legitimate business purpose? Yes. Was selling that data to shape election outcomes ethical? Arguably not. The issue lay in how the data was collected and whether users had truly consented to its use.

AI Ethics is a global debate. Some nations have responded through law. Others, through principles. Australia chose the latter. Instead of creating AI-specific legislation, the government introduced the Voluntary AI Safety Standard (the guiding principles for responsible AI design and deployment).[5] These cover accountability, transparency, testing, and explainability, but carry no legal force. There are no penalties for non-compliance or regulators to enforce them. In essence, they serve as a moral compass rather than a rulebook, relying on organizations to act ethically because it is right, not because it is required.

While ethics is sometimes dismissed as a "soft" discipline without real weight, nothing could be further from the truth. It is a forward-looking form of risk prevention.[6] Boards that act ethically often avert reputational crises, regulatory reform, or costly litigation before they materialize. Those that ignore ethics may meet the letter of compliance in the short term but lose legitimacy and trust in the long term.

Wells Fargo is a vivid reminder of such ethics showcasing that ethics runs through the organization. When culture is competitive and stakes are high, an environment is created where ethics is eroded because it must. In 2016, Wells Fargo paid $185 million in fines for 'creating about 1.5 million unauthorized deposit and 623,000 credit card accounts in customers' names without their knowledge.' [7] The scandal quickly escalated to a congressional "please explain," placing Wells Fargo's leadership under direct public scrutiny. In its aftermath, the bank replaced its Chairman and refreshed the board, publishing a report titled Rebuilding Trust in 2017 to signal reform.[8] Financial consequences followed. Wells Fargo paid over $185 million in regulatory fines, settled class-action lawsuits for $142 million , and saw more than 1,200 employees exit the organization. The fallout did not end there.In 2020, the bank agreed to pay $3 billion to the U.S. Department of Justice and SEC to resolve civil and criminal investigations into its sales practices.

When push comes to shove, and ethics alone cannot sustain trust and commitment, law intervenes. This is evident in the EU's AI Act and Denmark's proposed Human Face Copyright Act, both of which translate ethical principles into enforceable governance.[9] Organizations operating in jurisdictions that allow a long ethical rope should take stock of the trust placed in them. Ignoring that responsibility risks turning the rope into a much shorter leash, one that is tightened by regulatory intervention, fines, and lasting reputational damage. The case of Cambridge Analytica illustrates that ethical lapses not only triggered legal consequences but ultimately can drive a company into bankruptcy.

Fundamentally, ethics provides the foundation for governance while the law catches up.[10] It provides the ground on which trust is built. And in the digital age, where technology often outpaces regulation, ethics has become a core element of fiduciary responsibility. Boards that see ethics as a soft discipline misunderstand its power. What seems intangible one day can harden overnight into financial loss, regulatory scrutiny, and reputational collapse. The real question for directors is not whether ethics matters, but whether the organization's ethical ethos is strong enough to guide decisions when the law is quiet.

Types of Ethical Considerations

ESG (Environment, Social and Governance)

Ethics asks whether an organization is doing the right thing. ESG, which stands for Environmental, Social, and Governance, translates that question into measurable practice.[11] It requires organizations to demonstrate not only what they are doing but how they are doing it, and whether those practices are sustainable, transparent, and socially responsible. In this sense, ESG is the practical expression of ethics.[12] It provides the frameworks and metrics that hold companies accountable not only to regulators and shareholders, but also to employees, communities, and the planet.

ESG was once treated as a side note, often confined to a glossy section at the back of an annual report. Today, it has become a central measure of whether an organization is building trust and resilience. Artificial intelligence makes this shift unavoidable. AI is changing how we live, work, and govern, and in doing so, it is also changing how we consume resources, manage risk, and define responsibility.[13] Its benefits are evident in medicine, logistics, and finance, but its costs are equally real. These costs are not limited to budgets or reputational harm. They extend into environmental and social domains where they can no longer be ignored.

Behind every algorithm lies a vast and very physical infrastructure. Warehouses filled with servers, rivers diverted for cooling, and power grids strained by machines running without pause all form the hidden backbone of digital life. Data centers are not an abstract idea, they are physical sites with ecological footprints that rival entire industries. In August 2025, the ABC reported that within a decade AI data centers could consume one quarter of Sydney's water supply.[14] This is a city already vulnerable to water restrictions.

The report reframed AI from a digital marvel into a local resource dilemma. Will water serve the people of Sydney, or will it be diverted to the machines? At the same time, China announced a thirty seven billion dollar mega data center project in Wuhu.[15] Officials openly argued that in the modern economy, data is more valuable than food or land. These signals tell us something profound. Artificial intelligence is not just consuming electricity, it is competing for the same essential resources that sustain life.

The scale is astonishing. By 2024 there were almost twelve thousand operational data centers worldwide, with more than five thousand in the United States alone.[16] Some are small edge facilities, but others are sprawling hyper-scale sites that cover millions of square feet, the size of entire cities. In 2023, American data centers consumed one hundred and forty seven terawatt hours of electricity, roughly three percent of national usage.[17] Globally, demand is rising at a pace that regulators struggle to measure. Goldman Sachs has projected that AI alone could drive a one hundred and sixty percent increase in global power demand within the next few years.[18]

An additional power consumption feature of data centers is cooling. Cooling accounts for up to forty percent of data center energy use.[19] High performance chips generate such heat that operators now experiment with liquid immersion, underground chambers, and even ocean-based cooling systems. These innovations delay but do not erase the trade offs. Every liter of water diverted to a data center is water denied to households, agriculture, and ecosystems.

In 2024, OpenAI submitted a formal response to U.S. authorities proposing the construction of 5 gigawatt data centers in multiple states.[20] Their submission argued that a single facility of that scale could support or create around 40,000 jobs and contribute between seventeen and twenty billion dollars to local GDP. The ambition is astounding. I calculated that a center of this size would demand an energy load equivalent to five nuclear reactors, not for cities or communities, but to cool machines that serve artificial intelligence. The proposal brings the ethical dilemma into sharp focus. Should societies and investors reward scale at any cost, or should ambition be weighed against environmental restraint?

What is framed as progress may also represent a boundary test. At what point does technological ambition override sustainability? When an organization openly seeks to build infrastructure that strains natural resources on this scale, it exposes its values as much as its business plan. The question is no longer whether AI is transformative, but whether pursuing growth on such terms aligns with long-term trust, resilience, and social license.

Mega data centers deliver economies of scale, but at extraordinary environmental and social cost. Micro and hybrid centers may be more expensive and complex, yet they align more closely with resilience, equity, and sustainability.[21] The decision between these models is not only a financial calculation. It is a statement of corporate ethos.

The environmental impact goes further. A 2024 investigation by The Guardian suggested that emissions from major technology firms, including Google, Microsoft, Meta, and Apple, may be up to six hundred and sixty percent higher than officially reported. In Australia, projections show that by 2030 data centers and AI workloads could consume between eight and fifteen percent of national electricity supply.[22] This positions artificial intelligence as a central actor in future climate debates, not as a background concern.

Table 1: Mega vs Micro. A Question of Corporate Ethos

Model	Pros	Cons
Mega Data Centers	Economies of scale (lower cost per compute unit)	Enormous environmental footprint (electricity, water, land)
	Centralized talent, infrastructure, and security	Single points of failure (grid stress, natural disasters, cyber risk)
	Easier for governments and investors to quantify economic benefits	Seen as prioritizing profit and speed over sustainability and equity
		Social license risk: community backlash over "water for machines"
Micro / Hybrid Data Centers	Locally powered, often by renewables (solar, wind, geothermal)	Higher costs per compute unit compared to hyper-scale
	Reduced latency, critical for real-time AI and edge applications	More complex management and orchestration across sites
	Distributed and resilient, failure in one does not collapse network	Some frontier-scale AI models still require centralized clusters
	Easier integration with communities (reuse of waste heat, etc.)	
	Scalable in smaller increments, avoiding "all or nothing" dilemmas	

These facts cannot be treated as externalities. ESG is no longer a marketing exercise.[23] It is risk management, stakeholder engagement, and fiduciary duty combined. When directors ignore the environmental or social consequences of AI they risk breaching their duty to act in the long term interests of the company and its stakeholders. Shareholders and regulators alike now expect environmental and social factors to be integrated into governance decisions, not added as an afterthought.[24]

The Senate in Australia has already called for oversight of AI's ecological footprint,[25] while investors across Europe and North America demand disclosures that go beyond financial risk to include water, carbon, and social equity. The message is clear. ESG is the measurable backbone of ethical AI governance.

The challenge is that the footprint of artificial intelligence goes beyond electricity and water. It extends into rare earth mining, where the minerals required for processors carry human and ecological costs often borne by vulnerable communities.[26]

It is visible in dark data, the vast collections of unused information that waste storage capacity and consume energy without generating value.[28] It appears in the competition between cryptocurrency and AI, both of which fight for the same chips and power grids.

Even renewable energy brings dilemmas, since solar panels and wind turbines cannot yet be recycled at scale and may themselves become a source of future waste. Boards must therefore widen their focus.[29] The ESG cost of AI is not just operational. It is embedded in supply chains, labor practices, and end of life systems.

Recognizing the challenge is only the first step. Boards should also develop solutions. There are practical choices that can be made. Some organizations are turning to micro or hybrid data centers, combining on site facilities with off site ones powered by local renewables. Others are adopting edge processing so that not every transaction or sensor feed must be stored in a central cloud.

Data minimization is another solution. While some data must be held for legal purpose, other data may not hold the same value. As such, board should ask whether every click and camera feed, every form or transcript should be stored indefinitely. Arguably, tighter retention rules can save both money and emissions.

Procurement is equally a ‘power’ lever. Directors can demand that suppliers provide credible carbon and water reporting, and tie contracts to sustainability performance. In regions like Australia, investment in water recycling and closed loop cooling is becoming essential. Transparency in disclosures must be non negotiable. AI’s environmental and social impact belongs at the center of ESG reporting, not hidden in footnotes.

These steps are important because ESG is about results, not aspiration. The environmental and social costs of AI adoption are not abstract. They are measured in megawatts, megaliters, and the trust of communities. By embedding sustainability choices in the way AI is built and operated, organizations can reduce their footprint while preserving social license. Boards must ask: what is the cost of this initiative, can we meet the same objective with less impact, and how will we explain our trade offs to stakeholders?

The social dimension is equally vital. The ‘S’ in ESG reminds us that AI can either reinforce or undermine equity and dignity. Recruitment systems that disadvantage women, credit tools that penalize minorities, or predictive policing that targets vulnerable communities are not technical errors.[30] They are social failures with ethical and reputational costs. Boards must ensure management safeguards against bias and protects trust. Social sustainability is as important as environmental sustainability.

The ‘G’ is what ties everything together. Governance means structures of accountability, transparency, and oversight. It means directors asking the hard questions before crises arrive and refusing to accept vague reassurances. A governance failure is not just a missed disclosure. It is the absence of guardrails that prevent harm. The record of corporate scandals shows that when governance fails, law eventually intervenes. Boards that treat ESG as a reporting obligation rather than a guiding principle invite regulation, fines, and reputational collapse.

ESG expectations are set to become even more rigorous. Mandatory disclosures of AI's environmental and social impact are likely to become standard. Climate negotiations may begin to count AI as a formal emissions driver. Geopolitical tensions over rare earth minerals, energy grids, and water will only increase. Investor activism will grow stronger, pushing companies to demonstrate not just compliance but leadership in responsible AI. ESG is not optional. It is the arena where ethical intent is tested against measurable outcomes.

Ethics defines intent, but ESG holds organizations accountable for results. In the context of artificial intelligence this means directors cannot outsource responsibility to technologists or regulators. The choices about how AI is built, powered, and deployed are governance choices with ethical impact. By embedding ESG into AI strategy now, boards can align innovation with responsibility, protect trust, and secure long term resilience. The age of AI demands more than compliance. It demands ESG leadership grounded in ethics.

Data & AI (The Ethical Compass)

Artificial intelligence presents a paradox. On one hand, it offers extraordinary opportunities for efficiency, innovation, and economic growth. On the other, it amplifies risks in ways that law has not yet caught up to. In 2023, I prepared a submission to the Australian Federal Parliament titled Legal Considerations for a Safe and Responsible AI

Future in Australia. That paper focused on data quality, privacy, and the legal implications of algorithms in courts.3[1] Yet as I wrote it became clear that every one of those questions was also an ethical one. They asked not only what the law allowed, but what was right.

Boards must recognize that compliance cannot be the endpoint. In many cases the law remains silent, or slow to adapt, while technology reshapes human and organizational life in real time. What boards need is not just a checklist but a compass: a way to navigate questions of fairness, transparency, accountability, privacy, and human autonomy when regulation has not yet provided clear direction.

AI models are built on data, and data carries history. If past hiring patterns favored men over women, an AI-driven recruitment tool may simply learn to repeat that outcome. If policing records disproportionately target certain ethnic communities, predictive crime models will reinforce that bias rather than correct it. This has already happened in practice.

In the United States, the COMPAS algorithm was widely used in courts to predict the likelihood of re-offending.[32] Investigations found that it systematically gave higher risk scores to Black defendants than to white defendants with similar histories. Although marketed as objective, the system reflected the prejudices baked into its training data.
Such failures are not confined to America. In Australia, concerns have been raised over facial recognition technologies trialed in retail spaces by companies such as Woolworths and Bunnings.[33] Investigations by the Office of the Australian Information Commissioner revealed that biometric data was being collected without meaningful consent, with the risk of unfair profiling of shoppers.

What is most noticeable is that in each case the technology did what it was designed to do, yet the ethical outcomes were deeply questionable. Compliance with the letter of the law was not enough to prevent harm. Boards should ask whether their systems have been tested for fairness across different demographics, and whether processes exist for redress when bias is uncovered.

The second challenge arises from the complexity of modern AI. Some of the most accurate models are also the most opaque. Neural networks with billions of parameters can generate results that even their designers cannot fully explain. This creates what is often called the "black box" problem. For consumer services such as movie recommendations opacity may not matter, but in criminal justice, healthcare, and financial services the inability to provide reasons undermines trust.

The risk is already visible. In the Netherlands, a welfare fraud detection algorithm was deployed without sufficient transparency.[34] When challenged, officials struggled to explain how the system reached its conclusions. The program was removed when it was identified that it violated human rights.

When decisions affect people's lives, explanations are essential. Boards should confirm that systems are designed to be transparent and that human oversight exists when explanations are not possible. Without this, even accurate systems can fail the ethical test because accountability is missing.

Accountability is the cornerstone of trust. When an AI system makes a decision that causes harm, who is responsible? Is it the programmer who coded the model, the manager who approved deployment, or the board that endorsed its use? At present, liability is often contested, and courts struggle to assign blame in cases where AI systems malfunction or misclassify.

Yet ethics demands clarity where law hesitates. The tragic case of a teenager in the USA, who followed harmful recommendations from an AI-powered chatbot, illustrates the human cost of accountability gaps.[35] No single person set out to cause harm, yet the system's design, deployment, and oversight all contributed. The public rightly asked: who allowed this system into the world, and why were safeguards absent? Boards must ensure that clear lines of accountability are established before deployment, with escalation paths and human oversight explicitly defined. Machines may calculate, but ultimate responsibility must remain with people.

AI is powered by data, and the appetite for data has become insatiable. Every online search, shopping preference, and location ping can be harvested, stored, and analyzed. While the justification is often framed as efficiency or personalization, the result is that individuals are increasingly tracked in ways they did not anticipate or consent to.

The consequences are evidence. Clearview AI, a facial recognition company, scraped billions of images from social media platforms without permission, creating a global biometric database.[36] Law enforcement agencies in multiple countries accessed the system, often without legal basis or transparency. Even when regulators intervened, the damage was done. People's most intimate data had been turned into a surveillance tool without their knowledge.

Australia ha faced similar dilemmas. Privacy laws remain weaker than those in Europe, where the General Data Protection Regulation provides stronger protections. As AI advances, the risks will only intensify. Boards cannot wait for regulators to close these gaps. They must take a proactive stance, asking what data is being collected, why it is being stored, and whether individuals have been meaningfully informed. If those answers are unclear, then the company is already on the wrong side of the ethical line, even if it has not yet crossed a legal one.

Perhaps the most subtle tension is the effect of AI on human judgment. The more we rely on machines, the less we exercise our own capacity to decide. Already there are signs of "deskilling," where professionals defer to AI outputs without critical review. In one widely reported case in the United States, lawyers submitted a court filing prepared by ChatGPT that cited six cases which did not exist. The lawyers had not checked the references because they trusted the system's authority. Similarly, in predictive policing and credit scoring, human reviewers often default to algorithmic outputs rather than applying independent judgment.

The risk is not only professional embarrassment or inefficiency, but a deeper erosion of human autonomy. When organizations and individuals outsource judgment to machines, they weaken their own capacity to act responsibly. Efficiency is gained, but independence is lost. Boards must ask which decisions must remain human, regardless of how capable machines become. They must set clear boundaries for when AI may assist, and when humans must decide. Without these boundaries, autonomy risks being traded away piece by piece until dependence on machines is irreversible.

These tensions are not hypothetical. They are already shaping trust, regulation, and corporate reputation. Bias, opacity, accountability gaps, surveillance practices, and over-reliance on machines all erode confidence when left unchecked. Law may eventually respond, but ethics demands that boards act sooner. Directors cannot hide behind the excuse that "the system decided." They must ask whether their organization's choices in building and deploying AI are fair, transparent, accountable, respectful of privacy, and protective of human autonomy.

Legal, But Is It Right?

Law provides the rules of the road. It tells us which side of the road to drive on, how fast we can go, and when to stop at the lights. Without those rules there would be chaos, collisions, and no common understanding of order. From a objective view, rules of the road are only part of a journey. Ethics is everything that happens in between. It is the choice of which road to take, whether to stop and help someone at the roadside, and how much risk you are prepared to accept as you travel. Law sets the minimum requirements for safety and order. Ethics asks the deeper questions about direction, responsibility, and intent.

In the world of artificial intelligence, that difference is critical. Many of the most controversial uses of AI sit firmly within legal boundaries, yet they raise uncomfortable ethical questions. Clearview AI built a global facial recognition database by scraping billions of images from social media and CCTV feeds.[37] In some jurisdictions, this did not technically break the law, since the images were publicly accessible and consent was buried deep in terms and conditions. But the ethical cost was enormous. People discovered their photos were being used for purposes they had never agreed to, and governments eventually moved to ban or restrict the practice. The legality of the action did not shield the company from reputational collapse or regulatory backlash.

Another example lies in recommender systems used by social media platforms. Algorithms designed to maximize engagement often push extreme content because outrage and shock keep users online longer. From a legal perspective, this is simply content curation, protected in many jurisdictions by broad free speech provisions. Yet ethically it has fueled polarization, spread disinformation, and harmed public health by amplifying conspiracy theories. The companies involved could argue they complied with the law, but compliance did not protect them from being accused of undermining democracy itself.

Australia offers its own example of how law and ethics can sometimes be brought into the 'spotlight'. The SafeWA App was introduced during the pandemic as a contact-tracing tool, but data collected through it was later accessed by police for unrelated criminal investigations.[38] At the time, this was not unlawful, but it was widely condemned as an ethical breach of public trust.

Citizens who believed they were contributing to health and safety discovered their data had been repurposed without consent. The result was more than reputational damage. It created a chilling effect on public willingness to share data in the future.

This is not a philosophical exercise. It goes to the heart of fiduciary duty. Shareholders may tolerate aggressive strategies, but they do not forgive reputational implosions. Communities may embrace innovation, but they will not trust companies that exploit loopholes to act against public expectations. The sharper question is not "Can we?" but "Should we?" Would this decision stand up to public scrutiny if explained in plain language? Would customers feel respected or betrayed?

The law may tell you the speed limit, but it does not tell you when to slow down, or when to choose a safer road even if it takes longer. Ethics is that compass. It ensures the journey is not only legal, but also justifiable, trusted, and safe.

Reference

1. Paula Baron and Lillian Corbin, Ethics and Legal Professionalism in Australia (Oxford University Press, 2018).
2. Australian Constitution s 51.
3. Carole Cadwalladr and Emma Graham-Harrison, 'Revealed: 50 Million Facebook Profiles Harvested for Cambridge Analytica in Major Data Breach' The Guardian (18 March 2018) https://www.theguardian.com/news/2018/mar/17/cambridge-analytica-facebook-influence-us-election.
4. Christopher Wylie, 'Cambridge Analytica: How Did It Turn Clicks into Votes?' The Guardian (6 May 2018) https://www.theguardian.com/news/2018/may/06/cambridge-analytica-how-turn-clicks-into-votes-christopher-wylie.
5. Department of Industry, Science and Resources (Cth), Voluntary AI Safety Standard (Australian Government, 2024) https://www.industry.gov.au/publications/voluntary-ai-safety-standard.
6. 'Ethics vs Laws: 5 Key Differences Experts Want You to Know in 2024' (Infonetica, 2024) https://www.infonetica.net/articles/ethics-vs-laws.
7. Bill Chappell, 'Wells Fargo Fined $185 Million Over Creation of Fake Accounts' NPR (2016) https://www.npr.org/sections/thetwo-way/2016/09/08/493130449/wells-fargo-to-pay-around-190-million-over-fake-accounts-that-sparked-bonuses.
8. Wells Fargo & Company, 2017 Annual Report: Rebuilding Trust (Wells Fargo & Company, 2017).
9. European Parliament and Council of the European Union, Regulation (EU) 2024/1689 Laying Down Harmonised Rules on Artificial Intelligence (AI Act) (Official Journal of the European Union, 2024); Cat Woods, 'Denmark Proposes Copyright Laws to Protect Against Deepfakes' Law Society Journal (16 July 2025) https://lsj.com.au/articles/denmark-proposes-copyright-laws-to-protect-against-deepfakes/.
10. Ethical and Legal Practice, ATrain Education (Web Page) https://www.atrainceu.com/content/5-ethical-and-legal-practice.
11. Ethics Centre, 'ESG Is Not Just About Ticking Boxes – It's About Earning the Public's Trust' (Ethics Centre, 2024) https://ethics.org.au/esg-is-not-just-about-ticking-boxes-its-about-earning-the-publics-trust/.

12. Paula Baron and Lillian Corbin, Ethics and Legal Professionalism in Australia (3rd ed, Oxford University Press, 2023).
13. Arthur A Thompson Jr, Margaret A Peteraf and John E Gamble, Crafting and Executing Strategy: Concepts and Readings (23rd ed, McGraw-Hill Education, 2021).
14. Emily Jane Smith, 'AI Data Centres Could Use a Quarter of Sydney's Drinking Water by 2035' ABC News (27 August 2025) https://www.abc.net.au/news/2025-08-27/ai-to-take-up-one-quarter-of-sydney-water-in-a-decade/105700928.
15. Che Pan and Brenda Goh, 'China Spotlights Major Data Centre Project Using Domestic Chips' Reuters (17 September 2025) https://www.reuters.com/business/media-telecom/china-spotlights-major-data-centre-project-using-domestic-chips-2025-09-17/.
16. Brightlio Blog, '255 Data Center Stats (September 2025)' (29 September 2025) https://brightlio.com/data-center-stats/.
17. Sebastian Moss, 'DOE: Data Centers Consumed 4.4% of US Power in 2023, Could Hit 12% by 2028' Data Center Dynamics (2025) https://www.datacenterdynamics.com/en/news/doe-data-centers-consumed-44-of-us-power-in-2023-could-hit-12-by-2028/.
18. Goldman Sachs, 'AI Is Poised to Drive a 160% Increase in Data Center Power Demand' (Goldman Sachs Insights, 2024) https://www.goldmansachs.com/insights/articles/ai-poised-to-drive-160-increase-in-data-center-power-demand/.
19. Ibid (Data Center Dynamics duplicate for cross-reference).
20. OpenAI, Response to the NTIA Request for Comments on Bolstering Data Center Growth, Resilience, and Security (2024) https://cdn.openai.com/global-affairs/final-ntia-request-for-comments-on-bolstering-data-center-growth-resilience-and-security.pdf.
21. Shad Secrist, 'Understanding the Differences Between 5 Common Types of Data Centers' Data Center Frontier (18 May 2022) https://www.datacenterfrontier.com/sponsored/article/11427373/belden-understanding-the-differences-between-5-common-types-of-data-centers.

22. Isabel O'Brien, 'Data Center Emissions Probably 662% Higher than Big Tech Claims: Can It Keep Up the Ruse?' The Guardian (15 September 2024) https://www.theguardian.com/technology/2024/sep/15/data-center-gas-emissions-tech.
23. The Ethics Centre, 'ESG Is Not Just About Ticking Boxes – It's About Earning the Public's Trust' (2024) https://ethics.org.au/esg-is-not-just-about-ticking-boxes-its-about-earning-the-publics-trust/.
24. Organisation for Economic Co-operation and Development (OECD), G20/OECD Principles of Corporate Governance (2023).
25. Parliament of Australia, Select Committee on Adopting Artificial Intelligence (AI), Final Report: Chapter 6 – Impacts of AI on the Environment (November 2024) https://www.aph.gov.au/Parliamentary_Business/Committees/Senate/Adopting_Artificial_Intelligence_AI/AdoptingAI/Report/Chapter_6_-_Impacts_of_AI_on_the_environment.
26. Data Center Dynamics, 'DOE: Data Centers Consumed 4.4% of US Power in 2023, Could Hit 12% by 2028' (2024) https://www.datacenterdynamics.com/en/news/doe-data-centers-consumed-44-of-us-power-in-2023-could-hit-12-by-2028/.
27. Björn Schembera and Juan M Durán, 'Dark Data as the New Challenge for Big Data Science and the Introduction of the Scientific Data Officer' (2020) 33 Philosophy & Technology 93 https://doi.org/10.1007/s13347-019-00346-x.
28. Ibid (repeat reference for cross-citation).
29. 'Recycling End-of-Life Solar Panels, Wind Turbines Is Big Waste Business' CNBC (13 May 2023) https://www.cnbc.com/2023/05/13/recycling-end-of-life-solar-panel-wind-turbine-is-big-waste-business.html.
30. Jeffrey Dastin, 'Amazon Scraps Secret AI Recruiting Tool That Showed Bias Against Women' Reuters (11 October 2018) https://www.reuters.com/article/amazon-ai-recruiting-bias-idUSKCN1MK08G.
31. Brenda Van Rensburg, Submission to the Australian Federal Parliament — Legal Considerations for a Safe and Responsible AI Future in Australia (Department of Industry, Science and Resources, 2023) https://consult.industry.gov.au/supporting-responsible-ai/submission/view/63.

32. Julia Angwin, Jeff Larson, Surya Mattu and Lauren Kirchner, 'Machine Bias: There's Software Used Across the Country to Predict Future Criminals. And It's Biased Against Blacks' ProPublica (23 May 2016) https://www.propublica.org/article/machine-bias-risk-assessments-in-criminal-sentencing.
33. Ange Lavoipierre, 'Bunnings Breached Privacy Laws by Using Facial Recognition on Customers, Commissioner Finds' ABC News (19 November 2024) https://www.abc.net.au/news/2024-11-19/oaic-investigation-into-bunnings-facial-recognition/104613700; James Vyver, 'Australian Retail Giants and Police Using Artificial Intelligence Software Auror to Catch Repeat Shoplifters' ABC News (10 June 2023) https://www.abc.net.au/news/2023-06-10/retail-stores-using-ai-auror-to-catch-shoplifters/102452744.
34. Melissa Heikkilä, 'Dutch Scandal Serves as a Warning for Europe over Risks of Using Algorithms' Politico (2022) https://www.politico.eu/article/dutch-scandal-serves-as-a-warning-for-europe-over-risks-of-using-algorithms/.
35. Angela Yang, Laura Jarrett and Fallon Gallagher, 'The Family of Teenager Who Died by Suicide Alleges OpenAI's ChatGPT Is to Blame' NBC News (2025) https://www.nbcnews.com/tech/tech-news/family-teenager-died-suicide-alleges-openais-chatgpt-blame-rcna226147.
36. Office of the Australian Information Commissioner, 'Clearview AI Breached Australians' Privacy' (OAIC, 2021) https://www.oaic.gov.au/news/media-centre/clearview-ai-breached-australians-privacy.
37. Office of the Australian Information Commissioner, 'Statement on Clearview AI' (OAIC, 3 November 2021) https://www.oaic.gov.au/news/media-centre/statement-on-clearview-ai.
38. 'Police Refused to Stop Accessing SafeWA App Data Despite Premier's Orders, Inquiry Told' ABC News (16 June 2021) https://www.abc.net.au/news/2021-06-16/police-refused-to-stop-accessing-safewa-app-data-premier-says/100218764.

EXECUTIVE SUMMARY (Part 3)

Part 3 explored the foundations of responsible AI oversight through governance, accountability, and ethics. Together, these principles provide the lens through which boards evaluate every AI decision.

Governance was examined across five dimensions: corporate, enterprise, IT, data, and AI. Each reinforces transparency, accountability, fairness, and responsibility. Corporate governance defines who holds authority and how it is exercised. Enterprise governance links performance with conformance, ensuring that growth aligns with trust. IT governance manages technological decision-making, data governance safeguards information integrity, and AI governance integrates ethical oversight into intelligent systems.

Accountability and fiduciary duty were discussed as cornerstones of trust. Directors must consider not only shareholder returns but also the wider impacts on employees, customers, and society. Fiduciary responsibility extends to reputation, resilience, and the ethical use of data and algorithms.

Ethics and ESG formed the moral compass of this discussion, challenging boards to move beyond compliance toward legitimacy. Governance must consider not only what is legal but what is right, as demonstrated by issues like surveillance, data scraping, and algorithmic bias.

Finally, the section explored the infrastructure supporting this ecosystem, from micro to mega data centres. Their environmental and operational implications highlight that governance extends beyond policy into the physical and digital systems that sustain AI.

Together, these elements ensure AI serves human values while advancing sustainable innovation.

KEY TAKE AWAYS

- Boards remain accountable for the decisions they make about AI, and that accountability extends beyond compliance to ethics and societal impact.
- Principle 3 of the Corporate Governance Principles reinforces the board's duty to act ethically and responsibly in the organization's best interests.
- Directors hold both a legal and ethical responsibility for how data is collected, managed, and used, including how AI systems draw on that data.
- The environmental implications of AI were explored through the contrast between mega and micro data centres, helping boards understand that technology choices also carry sustainability consequences.

PART 4

Implementing AI

“Implementing AI
starts with
Purpose & Strategy.
Without it is like
Sailing a boat
without a
Map or a compass.”

CHAPTER 14

Strategy, Purpose and Scale

Strategy starts with clarity of purpose. Why is AI being considered, what problem does it intends to solve? Without these questions, strategy becomes reactive and governance becomes a box-ticking exercise. Purpose sets the boundaries against which decisions should be tested while ensuring that adoption remains aligned with organizational goals and its values. When purpose drifts, so too does strategy and directors risk moving from innovation into negligence

Depending on the size and maturity of a business, implementing AI may be as simple as flicking a switch or as complex as a multi-year program to ensure every box is ticked. I discussed AI, models, as well as the considerations of security, risk, governance, and ethics. Collectively, these form the cornerstone of responsible AI adoption.

However, there is an increasing demand for board members to support the implementation of AI without fully understanding its impact on people, the business, and associated risks. A decision to implement AI without proper due care could expose directors to personal liability.

Under section 180(1) of the Corporations Act 2001 (Cth), directors must exercise their powers and discharge their duties with the care and diligence that a reasonable person would.[1] The decision in Australian Securities and Investments Commission v Hellicar [2012] HCA 17 illustrates this principle.[2] In that case, the directors of James Hardie Industries were found personally liable for approving a misleading ASX announcement about asbestos compensation. The High Court held that the directors had breached their duty of care and diligence, showing that boards cannot rely solely on management assurances and must actively verify the information on which their decisions are based.[3]

Developing Strategy

A organizational strategy must go beyond simply deciding to adopt AI. It should set out a clear vision of what the organization seeks to achieve, anchored in its broader goals, and define the purpose by explaining why AI is being adopted and which problems or opportunities it is intended to address.[4] From there, it should establish concrete objectives with measurable outcomes, underpinned by guiding principles such as fairness, transparency, and accountability.

A strong strategy also includes a roadmap that sequences investments, pilots, and scaling activities, while ensuring alignment with existing business plans, regulatory obligations, and the organization's risk appetite.[5] Together, these elements ensure that AI adoption is not a technical exercise but a business-driven program with clear direction and accountability.

When strategy goes wrong, the consequences can be significant. For example, in August 2025, Commonwealth Bank of Australia, fired 45 staff thinking AI could do their job better.[6] Not only did the system under-deliver, forcing the bank to attempt to rehire staff, but the approach may not have met national AI standards. For a major bank, falling short of baseline standards is not a small misstep.

Without full disclosure, CBA could argue that their implementation of AI was aligned with strategy. However, despite operating under the Voluntary AI Safety Standard, the bank failed to create trust with users or provide confidence that AI was being deployed safely and responsibly.[7] Additionally, they did not have a process for people impacted by AI Systems. The organization utilized its staff to train the AI, dismissed them once the system went live, and only after its failure attempted to re-engage the same employees. This chain of events demonstrates poor governance and a lack of procedural fairness, a position that would likely not withstand scrutiny in a court of law.

A typical business strategy generally includes a clear vision and mission, defined strategic objectives, resource allocation, market analysis, an implementation plan, and strong leadership.[8] However, the key distinction between this and an AI implementation strategy is that the former focuses on the overall business, while the latter is specific to the adoption and oversight of AI.[9] While a generic business strategy framework provides a useful foundation, an AI implementation strategy must go further. It should include a clearly defined AI purpose, a comprehensive risk assessment supported by a strong governance framework, and a structured approach to change management.

The consequences of poor strategic execution can be just as damaging as poor planning. In 2024, a former New South Wales government contractor uploaded personal information of flood-relief victims into ChatGPT while trying to streamline reporting.[10] At a strategic level, the state's digital transformation agenda was sound . It aimed to modernize service delivery and improve response times through artificial intelligence. However, the lack of execution discipline turned an innovative strategy into a reputational setback. Sensitive data were entered into a public AI tool with no privacy controls, undermining public confidence in a program that was meant to strengthen it.

This incident illustrates that good strategy alone is not enough. The state had a clear vision, measurable objectives, and a legitimate purpose, yet failed to translate these into effective controls, training, and accountability. The strategy aligned with policy goals but faltered in implementation because human factors, data governance, and operational risk were not fully considered.

The strategy must extend beyond purpose statements and innovation headlines. It needs to anticipate how people will act under pressure, how technology will be used in practice, and how unintended consequences will be managed. In strategy, as in ethics, intent matters, but execution defines success.

Weak oversight has tangible consequences. In 2025, Deloitte Australia was required to refund part of a government contract after delivering a 237-page report that contained numerous AI-generated errors.[11] The report, commissioned to review the welfare system's IT and compliance framework, included fabricated court quotations, references to non-existent academic papers, and misattributed sources.

The project had a clear strategic purpose and appeared sound on paper. However, the absence of strong validation, human review, and provenance checks turned a credible engagement into a reputational and financial setback. This incident illustrates how the lack of oversight in AI-assisted work can undermine trust and expose even established organizations to public and regulatory scrutiny.

Firm vision and contractual purpose were clear, but the failure was in process, quality, and governance. What should have been a flagship example of AI-enabled consulting became a public cautionary tale about hallucinations, trust, and accountability. Replacing humans with AI can seem like a logical step, offering lower costs, faster output, and round-the-clock availability. However, the human consequences often outweigh the short-term gains.

In 2025, several global companies across sectors from finance to media began cutting staff while publicizing their adoption of generative AI.[12] Some replaced entire departments only to find productivity decline, errors multiply, and brand reputation suffer.[13] Technology can scale efficiency, but it cannot replace accountability or culture. When a strategy treats people as cost centers rather than assets, it risks hollowing out the very competence that innovation depends on.

IKEA demonstrated how strategy can go right. When it introduced its AI bot "Billie" in 2023 to handle routine customer queries, it began with a clear purpose: free up staff to focus on higher-value work.[14] Instead of neglecting the people element, IKEA built its roadmap around them, retraining thousands of call-center employees as interior design advisors.

This aligned AI adoption with the company's broader vision of enhancing customer experience while creating opportunities for its workforce. Strong leadership and communication reinforced trust, turning potential disruption into a story of growth and transformation.

The effectiveness of any AI strategy depends on how well its core elements are applied: vision, purpose, objectives, governance, and people. When organizations chase speed, cost savings, or competitive optics, they invite ethical, governance, and reputational risks. When they anchor AI adoption in trust, fairness, and accountability, the technology becomes an enabler of long-term value. The test of a good AI strategy is not whether it keeps pace with competitors, but whether it strengthens the organization's integrity and resilience while delivering on its mission.

Boards often lose visibility when the move from strategy to execution happens. Implementation is where vision collides with cost, culture, and operational reality, and where oversight matters most.

Strategic Drift

Drift doesn't announce itself as risk in the beginning. It enters the boardroom as opportunity; as a promise of growth, efficiency, or market advantage.[15] Directors ask, "What more can we do with the data we already have?" Those questions cascade down through the organization, managers push for answers, and data teams confirm what is technically possible. By the time the results cycle back up, the insights have already been folded into strategy.

That opportunity always begins with data.[16] In the early days, collection was straightforward and largely innocent: names, addresses, phone numbers. This information was needed to send an invoice, deliver a product, or mail marketing material. The purpose was clear, and consent was usually assumed. But as the digital environment matured, so did the appetite to collect more. Every transaction, every interaction, every click became a data-point.

When margins are thin, and small percentages make the difference between profit and loss, the temptation to stretch data beyond its original purpose becomes hard to resist. Businesses begin to reason. If a customer is willing to interact with us, then the data from that interaction belongs to us. This viewpoint overshadows the foundation of trust on which those interactions took place. Customers engaged with a service expecting delivery of that service, not for their every digital movement recorded and mined for competitive gain.

AI's capability is exponential. Depending on the model and the practice, it can uncover patterns that even the trained human eye would miss. The challenge is no longer simply about data sovereignty, but about connecting systems and servers so that AI can access, compute, and deliver “intelligence,” giving a business a leading edge.

The impact of strategic drift depends on how far the organisation moves from its original intent before trust is eroded, or regulators and the market push back. Sometimes drift does not come from internal use of data at all. It arises when boards decide to share data with third parties who apply their own tools and analytics.[17] Outsourcing the complexity of pattern recognition or credit scoring to a specialist provider, can seem attractive. But it carries the same risks if the legal basis for sharing is weak.

A recent example of 'drift' is Unión de Crédito para la Financiación Mobiliaria e Inmobiliaria EFCD, which was fined €200,000 for forwarding customer data to a credit scoring system without a proper legal basis.[18] To the business, it may have looked like a logical extension of customer information to support lending decisions. To regulators, it was an unlawful repurpose of personal data.

As such, responsibility does not end at the boundary of the organization. Whether data is used internally or shared externally, it remains subject to the same standards of purpose, fairness, and legal justification. Delegating analysis does not delegate accountability. Boards must be prepared to answer not only for how data is used within their walls, but also for how it is processed by every partner and provider acting on their behalf.

Resource and Capability Oversight

Resourcing and capability is another key cornerstone of Implementing AI. Purpose and strategy set the direction, but delivery depends on whether the organization has the means to make it real.[19] Governance may provide the guardrails, but those guardrails only matter if there are people and resources inside them. A well-written policy means little if staff do not have the skills to apply it or if there is no budget to sustain it.

AI adoption is not just about hiring data scientists and engineers.[20]The organization also needs risk officers, compliance professionals, legal experts, ethicists, and sometimes social scientists to assess fairness and bias. This mix ensures that AI is not only built, but built responsibly. Risk assessments are a good test case.[21] If they are delegated to junior staff without the right experience, the results can be misleading. A poor assessment gives the business a false sense of security and will not hold up under regulatory or legal scrutiny.

Beyond hiring or developing staff, training and awareness must be prioritized.[22] This should cover understanding AI capability, knowing when human judgment is still required, and how to use tools responsibly. Training should be tailored for different groups. Technical staff need deep knowledge. Frontline employees need practical guidance. Leaders need to understand how AI changes decision-making. Without this broad base of awareness, even well-designed systems can be undermined by poor practice or misplaced trust.

Funding is critical area. Strategy without funding is wishful thinking. AI requires upfront investment in tools, staff, and infrastructure.[23] It also requires ongoing funding for maintenance, compliance, and monitoring. Too often, organizations underestimate these long-term costs. When

AI is built internally, that means retraining models, cleaning data, and continuously monitoring for drift and bias.

When AI comes from a vendor, the organization may not be doing this work directly, but it must still understand the risks. Staff should be trained to recognize signs of drift or unexpected behavior and know how to raise concerns with third parties. Cybersecurity and compliance functions must also be resourced to match the scale of adoption.[24] Boards should therefore expect to see both initial and recurring costs set out clearly. The question is not only whether the organization can afford to buy or build AI, but whether it can afford to run it safely year after year.

AI also requires a strong foundation of infrastructure. Additional, leadership should confirm that leadership has considered the right mix of compute power, data storage, cloud contracts, and cybersecurity controls. These decisions are not only technical. They affect resilience, cost, and sustainability. AI systems consume large amounts of energy, and storing data at scale adds to the organization's carbon footprint.[25] Boards should therefore ask whether infrastructure aligns with their ESG commitments.

- Is renewable energy being used where possible?
- Are data centers efficient?
- Are retention policies stopping unnecessary accumulation?

There are different paths to building this foundation. Some organizations invest in micro data centers, which can offer flexibility, lower latency, and environmental benefits if powered by sustainable sources such as hydrogen batteries. Others rely on large third-party providers who can deliver capacity at scale. Each option comes with trade-offs. Owning infrastructure brings greater control but higher cost and maintenance responsibility. Outsourcing reduces capital expense but increases dependency and raises questions about security, privacy, and exit rights. Boards should weigh these trade-offs carefully and confirm that whichever path is chosen aligns with the organization's strategy, risk appetite, and long-term sustainability goals.

Most organizations will not develop AI entirely in-house. They will rely on external vendors, consultants, and partnerships with universities or industry bodies. These arrangements can add speed and expertise but also bring risk. Boards should ensure due diligence is conducted on third parties. Contracts must cover data ownership, liability for errors, regulatory compliance, and exit rights. If a vendor fails, who is responsible? If they go under, how quickly can the organization pivot?

Resource and capability oversight is about ensuring AI adoption is possible and sustainable.[26] People, funding, infrastructure, and partnerships all form part of the picture. The board's role is not to provide the answers but to ensure the right questions are asked and that executives deliver credible, tested plans. Without this, AI adoption risks becoming either a wasted investment or a liability. With it, AI can be implemented in a way that is safe, responsible, and aligned with the organization's long-term purpose.

Implementing AI Governance

With purpose defined, strategy agreed, and resources scoped, the next step is to establish governance foundations.[27] Governance is not an end in itself but the guardrails that keep AI adoption aligned with the organization's intent. Without these guardrails, even the best-resourced strategy can drift. Boards should expect to see the executive translate the strategy into policies, standards, and oversight structures that give effect to what has been agreed.

Practical models exist. The Australian Government's AI Standards set out ten guardrails that form a strong baseline.[28] They call for:

- accountability processes with clear ownership of AI;
- ongoing risk assessments;
- protection of data and system integrity;
- rigorous testing, meaningful human oversight;
- disclosure to users;
- processes for appeal;
- transparency across the AI supply chain;
- comprehensive record-keeping;
- and active stakeholder engagement.

The value of these guardrails lies not in the detail but in the discipline. They represent the minimum structure required to ensure AI is deployed safely, ethically, and with accountability.

Security and risk management are already well established. We have seen how AI adds new layers of complexity. Issues such as supply-chain vulnerabilities, data quality, and model drift are not technical footnotes. They are governance issues. Boards must focus on whether existing structures adequately identify and manage such risks.

AI policies and standards are key for any organization looking to implement AI.[29] Like most corporate policies, they follow a familiar structure, but with AI the focus must shift to issues such as ethical principles, acceptable use cases, data privacy, and human oversight. A well-designed AI policy sets boundaries, clarifies responsibilities, and provides a framework for monitoring compliance.[30] Standards bring those principles to life, establishing the processes, safeguards, and documentation needed to make sure AI systems are deployed responsibly.

These policies become most meaningful when seen in practice. AI is already used for predictive maintenance, safety monitoring, and optimizing extraction, in mining. Without clear rules on data quality, bias, and worker oversight, safety-critical systems could fail, putting lives at risk.[31] In financial services, AI drives credit scoring, fraud detection, and trading. Policies on fairness, transparency, and explainability are essential to prevent discrimination, avoid regulatory breaches, and protect reputation.[32] In insurance, AI is used for claims processing and risk profiling. Without safeguards on privacy, human review, and appeals, automated systems can unfairly deny claims or expose sensitive personal data.

The board's role is to endorse AI policies and standards that align with strategy and are fit for purpose. Such policies should be formally presented when amended and reviewed annually. They must include a clear risk framework, with directors satisfied that risks are properly identified and assigned. Boards should review risk assessments for all internally developed and vendor-supplied AI systems.[33] Even when a tool aligns with strategy, regulatory, reputational, and legal risks remain. It is the board's duty to ensure governance structures are strong enough to manage them.

One of the most common failures in governance is the assumption that AI oversight naturally sits within IT.[34] This misperception could stem from the idea that AI is a technology, and therefore the responsibility must rest with the CIO or CTO. That is governance drift. Technology may be the delivery mechanism, but AI adoption cuts across risk, ethics, compliance, and strategy. It is therefore a board issue, not just a management one.

When responsibility is shared by everyone, it often ends up belonging to no one. Someone must be clearly accountable for how AI is governed and monitored. It does not matter what the committee is called. What matters is that someone is watching, asking questions, and speaking up when something is wrong.

Governance must also make clear when an AI issue comes back to the board for review. Oversight is not a one-time approval. It requires live reporting and defined escalation points. Major incidents, regulatory notices, reputational harm, data breaches, or signs of algorithmic bias should never be buried in operational reports. Boards should expect to see these issues raised quickly, with a clear plan for how management will respond. Without red-flag triggers, risks remain hidden until they become crises.

In larger organizations, responsibility is often reinforced by dedicated Data and AI teams. These teams provide the technical expertise needed to manage data quality, model testing, bias monitoring, and compliance with internal standards. Their role is not to replace board oversight but to support it.[35] It is about ensuring management has the depth of capability to meet obligations, while giving the board the insight it needs to exercise accountability. Smaller organizations may rely on existing structures such as Audit and Risk, but the principle is the same: governance must be anchored, capability must be in place, and accountability must remain with the board.

Most AI will be implemented through third-parties.[36] This introduces a layer of dependency that must be actively governed. Boards should require evidence of due diligence on these providers, covering security, data handling, fairness testing, and regulatory compliance. Contracts must clearly set out who owns the data, who carries liability if something goes wrong, and what rights the organization has to exit.[37] Without this, organizations risk being locked into costly, opaque, or even non-compliant systems.

Governance is also dependable on culture as it on policy.[38] Frameworks and committees can set expectations, but it is people who make AI accountable in practice. Culture determines whether those frameworks are lived or ignored.

AI is not only a matter of governance; it is also part of an organization's broader ESG responsibility. Every model trained and every dataset stored carries an environmental cost. Large-scale models consume significant energy to train and operate, increasing both electricity use and carbon footprint. The more AI an organization deploys, the greater its environmental impact.

Data management adds another layer of complexity. In some jurisdictions, legal obligations restrict how long information can be retained or deleted. In Australia, for instance, the Archives Act and various Records Acts require certain government records to be preserved for 25 years or longer, in some cases permanently. [39] These requirements can limit the ability to minimize or erase data, creating a tension between regulatory compliance, environmental responsibility, and sustainable data practices.

This is where data governance becomes essential.[40] Boards should expect management to distinguish between data that must be retained for legal or regulatory reasons and data that provides little long-term value. Mandatory records such as contracts or personnel files sit in a different category from website cookies or click-stream logs, which may have short-term utility but little justification for decades of storage. Good governance ensures that retention schedules are clear, data is classified appropriately, and unnecessary accumulation is avoided.

Boards should therefore ask the same hard questions about AI as they do about other resource-intensive operations. What is the energy cost of our AI systems? How much data are we storing, and do we need it all? What steps are in place to reduce or offset our carbon footprint? These questions are not technical; they are governance. They link directly to sustainability commitments and the organization's license to operate.

Anchoring oversight within a committee, supported where necessary by dedicated Data and AI teams, and embedding ESG considerations ensures AI adoption is not pursued at the expense of responsibility. Governance must extend beyond risk and compliance to include sustainability. When boards link AI oversight with both accountability and environmental impact, they create a governance structure that is robust, visible, and aligned with long-term value.

Because ESG is central to modern governance, boards should also be asking how much AI is actually needed, how much data is truly necessary, and whether their current AI choices risk creating long-term dependency. Vendor lock-in can bind organizations to systems that are costly to run, difficult to exit, and environmentally intensive. These questions push strategy and governance closer together by ensuring that AI adoption is not only responsible today but remains adaptable and sustainable in the future.

Test, Measure and Scale

Pilots are where AI strategy moves from theory into practice, by implementing it into a test environment. They give organizations a controlled environment to test assumptions, see how systems behave, and learn what works before committing large investments.[41] This is the point where oversight becomes especially important. Pilots often start small, but if they are not structured properly, they can drift into production without ever being properly evaluated. That is when risks multiply.

A well-run pilot should have its own mini-strategy.[42] This is not a full-scale corporate plan, but it must still define boundaries and expectations.

- Is the pilot running in a sandbox environment or in live operations?
- Who has access to it, and under what conditions?
- What testing rules apply, and how will results be measured?

Without this clarity, a pilot risks becoming a technology experiment rather than a governance tool. Boards should expect to see a clear purpose, an agreed timeframe, and measurable outcomes before any pilot begins.

The design of a pilot matters because it sets the stage for scaling.[43] A pilot that runs in isolation, without proper guardrails, can give misleading results. A system that works well in a narrow, controlled scenario may behave very differently when scaled across an entire customer base.

Questions to ask are: what are the entry criteria for this pilot, and what are the exit criteria? Success should be defined in advance. Equally important, so should failure. If the pilot does not meet its objectives, does the organization have the discipline to stop? Ending a pilot is not a failure. It is a sign of good governance. It prevents resources from being wasted and ensures lessons are learned before harm is done.

Scaling is the next challenge. Moving from a small test to broad deployment is not simply a matter of turning on more servers or processing more data. It is about proving that the system can operate fairly, securely, and reliably under real-world conditions. For example, a fraud detection tool tested on a limited dataset might perform well in a pilot, but when rolled out, it could mis-classify legitimate transactions and frustrate customers. Boards should press for assurance that pilots are tested not just for technical accuracy but also for fairness, transparency, and alignment with organizational purpose.

The scaling process should also include checks on resourcing.[44] What works with a handful of engineers in a pilot may require dedicated teams, funding, and infrastructure at full scale. Boards should ask whether the organization has the people, processes, and partnerships in place to handle scale responsibly. Without that, scaling risks outpacing capability. Review is the part of the cycle that too many organizations treat as an afterthought. Once a pilot has been approved to scale, boards should not assume the job is done. AI systems are not static. They evolve as data changes, environments shift, and models drift. A system that was safe and effective yesterday may not remain so tomorrow. This is why review must be continuous. Boards should expect regular reporting on performance, risks, and impacts, with structured checkpoints to decide whether adjustments are needed.

Continuous review also creates the opportunity to capture lessons learned and drive improvement. It reveals what worked well, what did not, and how the next pilot can be designed more effectively. Feedback loops are essential because they connect operations back to strategy, ensuring that AI adoption evolves in a way that remains both disciplined and adaptable. Without review, mistakes repeat, risks compound, and strategy begins to drift.

A failed pilot is not merely a technical setback. It can expose governance weaknesses, trigger regulatory scrutiny, and erode stakeholder confidence. Conversely, a pilot that is well designed, closely monitored, and properly reviewed becomes a proving ground that strengthens confidence in responsible AI adoption.

Ultimately, pilots, scaling, and reviews are not technical matters alone. They are governance checkpoints. They decide whether AI adoption strengthens the organization's purpose or undermines it. Boards should therefore hold management to clear expectations. Every pilot must have a mini-strategy. Scaling must not outpace capability and reviews must be genuine checkpoints, not a box-ticking exercises.

Example of AI Implementation

Having explored the theory of AI strategy, governance, and risk, it is useful to see how these concepts would play out in the real world.[45] A well written strategy means little until it is tested under practical conditions. The following example illustrates how each stage of implementation, from intent to review, can either strengthen or undermine success. By stepping through a real-world trial of AI in a fast-food context, we can see how a sound strategic vision, if not matched by governance, resources, and control, can drift into operational and reputational failure. The case reminds us that strategy is not only about intent but also about execution, oversight, and the ability to stop when the costs begin to outweigh the gains.

AI on the Menu but Not Ready to Serve

Step 1 (**Strategic Intent**): A restaurant partnered with a technology company to trial an AI voice-ordering system in more than 100 drive-thru locations. The vision was straightforward. Speed up service, reduce labor costs, and enhance customer experience. On paper, the strategy aligned with both efficiency and customer satisfaction.

Step 2 (**Build Strategy**): The objectives were measurable. To reduce order time by a set percentage, lower staffing costs during peak hours, and create a seamless drive-thru experience. What the strategy failed to anticipate was the unpredictable reality of the drive-thru environment. Noise, accents, and overlapping conversations proved challenging. The AI misheard, misinterpreted, and sometimes duplicated orders in bizarre ways. Hundreds of phantom items, bacon added to ice cream, or random extras that frustrated customers and went viral online

Step 3 (**Governance Foundations**): The pilot also revealed weak governance. Human override mechanisms were limited, thresholds for invalid orders were poorly defined, and monitoring of errors lacked rigor. The fact that basic misorders made it into customer hands suggested that governance did not enforce rigorous testing criteria or operational controls before deployment. Eventually, the restaurant terminated the program, canceling the partnership after reputational and operational costs became too high.

Step 4 (**Resource & Capability**): The AI required more than voice recognition. It needed sophisticated audio filtering, contextual understanding, error recovery, and adaptation to edge cases such as customizations or cross-talk. It is unclear whether the restaurant or the technology company had the resources and expertise to meet these challenges at scale. The mismatch suggested that technical and human resources were underestimated. The infrastructure for noisy, real-world drive-thru conditions required resilience that was not planned for.

Step 5 (**Risk & Control Assurance**): Errors quickly spread on social media, fueling reputational risk. Control mechanisms to prevent or roll back mis-orders proved insufficient, and in many cases human oversight was absent or ineffective. What could have been minor operational glitches escalated into brand-damaging events because risks were not adequately managed.

Step 6 (**Pilot, Scale & Review**): By the middle of the year, the restaurant ended the experiment, withdrawing the system from over 100 stores. Scaling so broadly had turned a pilot into a de-facto expansion. Once the flaws were discovered, reversing the decision became enormously costly. Hardware had been installed, staff trained, contracts signed, and management attention consumed. Beyond direct financial loss, the event was a public relation's nightmare.

IKEA's AI Success Story

Let's looking at a strategy that was, from my point of view, executed perfectly. IKEA provides a powerful example, a case where technology was used to strengthen, not replace, human capability.[46] Rather than treating AI as a cost-cutting tool, the company approached it as an enabler of efficiency, creativity, and long-term value. The result was not only measurable financial success but also a demonstration of how purpose, planning, and people can align to deliver innovation responsibly. While some details were unavailable, I developed a close hypothesis to illustrate what a successful implementation might look like.

Step 1 (**Strategic Intent**): IKEA defined its intent as enhancing operational efficiency and customer experience, not reducing headcount. The vision was for AI to assist staff, optimize inventory and supply chain logistics, and improve responsiveness, all while keeping employees in place.

Step 2 (**Build Strategy**): The objectives were measurable. Boost sales or revenue by a specific percentage, reduce waste, improve turn around times, or increase customer satisfaction. The strategy was not centred on replacing staff swiftly, but augmenting human work with AI support.

Step 3 (**Governance Foundations**):Before deployment, IKEA (or its partner) would have codified guardrails by ensuring policy around acceptable AI use was established, implementing oversight committees, ensuring human override rules, include error thresholds, and include transparency requirements. These guardrails ensure technology remains accountable to human values.

Step 4 (**Resource & Capability**):Rather than assuming AI would just run, they invested in the right mix of infrastructure (data, compute, cloud), staff training, operational support, and maintenance. They matched their ambitions with capability so the AI enhancements could be sustained.

Step 5 (**Risk & Control Assurance**): Throughout deployment, IKEA monitored for model drift, fraudulent or biased patterns, system performance, and user feedback. They ensured mechanisms existed to rollback or adjust features quickly if negative outcomes emerged.

Step 6 (**Pilot, Scale & Review**):IKEA didn't rush to full deployment. It scaled stepwise, learning as it went. Because of that, they were able to scale without major breakdowns or backlash. The result: a reported US$1.4 billion revenue uplift without layoffs. Their success shows that AI can drive value with human inclusion, not at the expense of people.

When purpose, strategy, and governance align, AI does more than deliver results. It elevates people. IKEA's success was not luck but leadership in action. It showed that technology guided by human values can create progress that is both profitable and principled. This is what responsible AI looks like. Innovation that strengthens trust, sustains growth, and keeps humanity at the heart of every decision.

Reference

1. Corporations Act (Cth) 2001.
2. Australian Securities and Investments Commission v Meredith Hellicar & Ors [2012] HCA 17.
3. Ibid.
4. Arthur A Thompson Jr, Margaret A Peteraf and John E Gamble, Crafting and Executing Strategy: Concepts and Readings (23rd ed, McGraw-Hill Education, 2021).
5. G Russell Hubbard, Strategic Management: Thinking, Analysis and Action (Pearson Education, 2000).
6. Heath Parkes-Hupton, '"Shocked" Commonwealth Bank Worker Trained AI Replacement' News.com.au (2025) https://www.news.com.au/finance/work/careers/shocked-commonwealth-bank-worker-trained-ai-replacement/news-story/c6ac3be4be84e8977d8742cc3d37af85.
7. Department of Industry, Science and Resources (Cth), Voluntary AI Safety Standard (Australian Government, 2024) https://www.industry.gov.au/publications/voluntary-ai-safety-standard.
8. G Russell Hubbard, Strategic Management: Thinking, Analysis and Action (Pearson Education, 2000).
9. Jarrod Anderson, The Chief AI Officer's Handbook (Packt Publishing, 2025).
10. Frank Chung, "Deeply sorry'. NSW contractor uploaded 3000 flood victims' data to ChatGPT (2025, News)https://www.news.com.au/technology/online/internet/deeply-sorry-nsw-contractor-uploaded-3000-flood-victims-data-to-chatgpt/news-story/f8d0777ebea5400090be402b8314dede.
11. Nino Paoli, 'Deloitte Was Caught Using AI in $290,000 Report to Help the Australian Government Crack Down on Welfare After a Researcher Flagged Hallucinations' Fortune (7 October 2025) https://fortune.com/2025/10/07/deloitte-ai-australia-government-report-hallucinations-technology-290000-refund/.
12. Aaron Drapkin, 'These Companies Replaced Workers with AI — Here's What Happened Next' Tech.co (8 October 2024) https://tech.co/news/companies-replace-workers-with-ai.
13. Stephanie Chalmers, 'Commonwealth Bank backtracks on AI job cuts, apologises for "error" as chatbot lifts call volumes' (Web page, 21 August 2025) ABC News

https://www.abc.net.au/news/2025-08-21/cba-backtracks-on-ai-job-cuts-as-chatbot-lifts-call-volumes/105679492.
14. SevenFour Digital, 'How IKEA's Use of AI Resulted in a $1.4 Billion Revenue Uplift with No Layoffs' SevenFour Digital (2024) https://sevenfour.digital/how-ikeas-use-of-ai-resulted-in-a-1-4-billion-revenue-uplift-with-no-layoffs.
15. Elena Shkarupeta, Aleksei Stroeva and Andrey Sybachin, 'A Framework for Digital Development of Industrial Systems in the Strategic Drift to Industry 5.0' (2022) ResearchGate https://www.researchgate.net/publication/366604772_A_Framework_for_Digital_Development_of_Industrial_Systems_in_the_Strategic_Drift_to_Industry_50.
16. Ibid.
17. Arthur A Thompson Jr, Margaret A Peteraf and John E Gamble, Crafting and Executing Strategy: Concepts and Readings (23rd ed, McGraw-Hill Education, 2021).
18. Enforcement Tracker, 'Unión de Crédito para la Financiación Mobiliaria e Inmobiliaria EFCD – Spain – Fine: €200,000' (CNPD, 2023) https://www.enforcementtracker.com/ETid-2641.
19. Arthur A Thompson Jr, Margaret A Peteraf and John E Gamble, Crafting and Executing Strategy: Concepts and Readings (23rd ed, McGraw-Hill Education, 2021).
20. Jarrod Anderson, The Chief AI Officer's Handbook (Packt Publishing, 2025).
21. K Scott Griffith, The Leader's Guide to Managing Risk: A Proven Method to Build Resilience and Reliability (Harvard Business Review Press, 2014).
22. Yassine Maleh et al, IT Governance and Information Security: Guides, Standards, and Frameworks (CRC Press, 2022).
23. Arthur A Thompson Jr, Margaret A Peteraf and John E Gamble, Crafting and Executing Strategy: Concepts and Readings (23rd ed, McGraw-Hill Education, 2021).
24. Philip Weinzimer, Strategic IT Governance 2.0 (CRC Press, 2023).
25. Ibid.
26. Sebastian Moss, Former Google CEO suggests building data centers in remote locations in case of nation-state attacks to slow down AI (2025, Data Center Dynamics) https://www.datacenterdynamics.com/en/news/former-google-ceo-suggests-building-data-centers-in-remote-locations-in-case-of-nation-state-attacks-to-slow-down-ai/.

27. Organisation for Economic Co-operation and Development (OECD), G20/OECD Principles of Corporate Governance (2023).
28. Department of Industry, Science and Resources (Cth), Voluntary AI Safety Standard (Australian Government, 2024) https://www.industry.gov.au/publications/voluntary-ai-safety-standard.
29. Tom Petzold, AI in the Boardroom: The Essential Guide to Navigating Artificial Intelligence for Boards and Executives (Routledge, 2023).
30. Gregory Falco and Eric Rosenbach, Confronting Cyber Risk: An Embedded Endurance Strategy for Cybersecurity (Oxford University Press, 2021).
31. Tom Petzold, AI in the Boardroom: The Essential Guide to Navigating Artificial Intelligence for Boards and Executives (Routledge, 2023).
32. Department of Industry, Science and Resources (Cth), Voluntary AI Safety Standard (Australian Government, 2024) https://www.industry.gov.au/publications/voluntary-ai-safety-standard.
33. Ibid.
34. K Scott Griffith, The Leader's Guide to Managing Risk: A Proven Method to Build Resilience and Reliability (Harvard Business Review Press, 2014).
35. Elena Shkarupeta et al, 'A Framework for Digital Development of Industrial Systems in the Strategic Drift to Industry 5.0' (2022) ResearchGate https://www.researchgate.net/publication/366604772.
36. Jarrod Anderson, The Chief AI Officer's Handbook (Packt Publishing, 2025).
37. Robert Tricker, Corporate Governance: Principles, Policies and Practices (5th ed, Oxford University Press, 2023).
38. Gregory Falco and Eric Rosenbach, Confronting Cyber Risk: An Embedded Endurance Strategy for Cybersecurity (Oxford University Press, 2021).
39. Arthur A Thompson Jr, Margaret A Peteraf and John E Gamble, Crafting and Executing Strategy: Concepts and Readings (23rd ed, McGraw-Hill Education, 2021).
40. State Records Act 2000 (WA).
41. Anita Kibunguchy et al, Data Governance: The Definitive Guide (O'Reilly Media, 2023).
42. Tom Petzold, AI in the Boardroom: The Essential Guide to Navigating Artificial Intelligence for Boards and Executives (Routledge, 2023).

43. Department of Industry, Science and Resources (Cth), Voluntary AI Safety Standard (Australian Government, 2024) https://www.industry.gov.au/publications/voluntary-ai-safety-standard.
44. Philip Weinzimer, Strategic IT Governance 2.0 (CRC Press, 2023).
45. Graham Kenny, Strategic Factors: Develop and Measure Winning Strategy (President Press, 2001).
46. Final Example of a Well Execute Plan: SevenFour Digital: How IKEA's Use of AI Resulted in a $1.4 Billion Revenue Uplift with No Layoffs", SevenFour Digital (Reviewed July 2025) https://sevenfour.digital/how-ikeas-use-of-ai-resulted-in-a-1-4-billion-revenue-uplift-with-no-layoffs/.

EXECUTIVE SUMMARY (Part 4)

Implementing AI is not about switching on technology. It is a board responsibility that begins with purpose. Boards and executves must first ask why AI is being introduced and what problem it aims to solve. Without this clarity, strategy drifts, governance becomes procedural, and accountability weakens.

Earlier chapters outlined the foundations of governance, showing how AI governance can be implemented in practice. An effective AI strategy sets measurable objectives, defines risk appetite, and is grounded in fairness, transparency, and accountability. Outsourcing infrastructure or analytics does not transfer responsibility. Boards remain accountable for risk assessment, ethical use, and procedural fairness, regardless of who manages the data.

Resourcing and capability determine success. AI requires more than engineers. It needs compliance, ethics, risk, and leadership disciplines working together. Budgets must be sustainable, infrastructure choices must align with ESG commitments, and contracts must make accountability explicit. Boards should seek assurance that the organization can sustain and govern AI, not just implement it.

Pilots should be tested before they scale. They must have clear measures, outcomes, and exit criteria. Scaling should only occur once lessons are proven and capability is in place.

Implementing AI is not just a technical project. It is governance. The measure of success is not speed but sustainability. What matters is whether adoptions deliver long term value, strengthens accountability, and builds trust.

KEY TAKE AWAYS

- Implementing AI begins with strategy and vision. Once this foundation is clear, it can guide and support implementation.
- AI governance provides oversight to ensure the strategy remains aligned with purpose, risk appetite, and accountability.
- Resourcing and capability must be built into the strategy, not added later. Success depends on having the right people, skills, and infrastructure in place.
- Boards should know when to persevere and when to pivot. Clear criteria for success and failure help prevent wasted investment and reputational risk.
- Real-world examples showed both sides of implementation: rapid adoption without planning that failed, and strategic adoption that achieved measurable return on investment.

EXECUTIVE SUMMARY (The Book)

Artificial intelligence is no longer a distant concept. It is reshaping every boardroom decision, redefining accountability, and testing the limits of governance. The question is no longer if organizations will use AI, but whether they will govern it well.

On Board AI equips leaders to do exactly that. It cuts through the noise to show how boards can lead with purpose, protect trust, and capture opportunity without crossing ethical or regulatory lines. It gives directors the clarity to ask the right questions, challenge assumptions, and make informed choices in a world where automation and accountability now intersect.

The book begins by explaining the fundamentals of AI. It explores systems, models, and products, and why the differences matter. It shows how algorithms and data drive intelligence and how every decision, from design to deployment, carries consequence. It shows how algorithms and data power intelligence, and how each decision, from design to deployment, shapes results.

The book then moves into risk, governance, and ethics. It explains how to identify AI-specific risks such as model drift, data poisoning, and over-reliance, and how to mitigate them through the Big Four foundations of security: organizational readiness, network, cloud, and software. It also shows how fiduciary duty and Principle 3 of corporate governance extend to understanding how algorithms and data shape decisions, ensuring that these technologies are governed with ethical awareness and accountability for their environmental impact.

The finale is strategy and implementation. It outlines how to build an AI roadmap with clear purpose, measurable goals, and accountable leadership. Case studies illustrate the contrast between organizations that rushed to deploy AI and those that integrated it within their governance frameworks from the outset. The outcomes are rarely determined by technology alone.

This book is written for leaders who want to bridge technology and governance. It was created to connect the technical and the strategic, giving boards the insight to lead responsibly and technologists the assurance that their work is understood and supported.

A Note from the Author

I wrote this book because I've seen how many organizations struggle with AI and the decisions around it. My aim was to make it easier for boards and executives to understand what matters, avoid common mistakes, and make choices that build trust and accountability.

I hope it guides with asking better questions, making stronger decisions, and guiding AI with purpose.

PART 5

Reference Material

APPENDIX 1 -GEN AI SCORE CARD

This scorecard, which I adapted from several governance framework, is designed to evaluate the strategic, operational, and ethical exposure of generative AI (GenAI) initiatives. Each domain represents a core area of governance oversight. Directors should assess where the current or proposed AI use case sits (Low, Moderate, or High) based on its level of integration, data sensitivity, operational dependency, and regulatory exposure.

This only an example. You can change the domain and low-to-high description.

Domain	Low	Moderate	High	Board Question
Purpose & Use	GenAI used for training, research, or internal lab use	GenAI applied as a support tool, not core to business operations.	GenAI embedded in customer-facing services, decision-making, or critical operations.	Has management clearly defined why this tool is being used, and does that align with strategy and risk appetite?
Business Integration	GenAI outputs are not connected to business processes.	Outputs feed into non-critical processes	Outputs are integrated into core processes where errors could cause material impact.	If integrated, what safeguards ensure outputs are accurate, reviewed, and appropriately governed?
Confidential & Sensitive Data	Confidential & Sensitive Data No customer, employee, or company confidential data is involved.	Limited use of internal (non-sensitive) data; not retained for training.	Sensitive or confidential data is stored, processed, or used in training.	What controls prevent misuse, retention, or leakage of sensitive data?
Operational Resilience	Outage or downtime has no impact on operations.	Outage causes disruption but manual fallback is feasible at reasonable cost or time.	Outage would materially impact operations; fallback is impractical or too costly.	Has the vendor demonstrated a tested fallback or recovery plan?
Regulatory, Compliance & Reputation	Minimal likelihood of regulatory, compliance, or reputational exposure.	Some potential for obligations or scrutiny depending on use case.	High risk of regulatory breach, legal action, or reputational damage.	Can Legal or Compliance confirm defensibility under current or emerging law?

APPENDIX 2 -ISO SCORE CARD

This is ISO 23894:2023 scorecard I created. It is an example only. Please visit: https://www.iso.org/standard/77304.html for a concise copy.

Domain	Key Questions for Boards	Evidence Expected	Y	N
Leadership & Liability	Has the board formally endorsed AI risk management as part of governance?	Board minutes, policy, committee terms of reference		
Intergration	Are AI risks integrated into enterprise risk management (ERM) and risk register?	Risk register entries, ERM reports		
Desing & Implementation	Were risk assessments conducted before deployment with clear acceptance criteria?	Pre-deployment risk assessment docs, approval papers		
	Do systems address fairness, safety, privacy, explainability?	Testing reports, DPIAs, fairness audits		
Evaluation & Improvement	Are AI systems reviewed for drift, bias, and harmful outputs?	Post-deployment monitoring logs, board reporting		
	Is there a process for continuous improvement?	Incident reviews, update cycles		
Risk Management Process	Is there structured communication with stakeholders?	Consultation records, stakeholder logs		
	Have scope and risk thresholds been defined?	Scope docs, risk appetite statement		
	Are monitoring and reporting cycles in place?	Quarterly reports, escalation logs		
Annex A Objectives	Is accountability clearly assigned (RACI model)?	RACI chart, governance framework		
	Does the org have AI expertise (internal or external)?	Skills register, advisory panel evidence		
	Is training/test data validated for quality & bias?	Data quality checks, documentation		
	Has environmental impact of AI projects been assessed?	ESG reports, sustainability analysis		
	Are systems tested for robustness, safety, and security?	Red team results, pen-test reports		
	Can outputs be explained (explainability)?	Explainability statements, model cards		
Annex B Risk Sources	Has complexity/interconnection risk been considered?	System maps, dependency logs		
	Is there clarity on level of automation and human oversight?	Policy on human-in-loop, escalation protocols		
	Have ML-specific risks (poisoning, overfitting, adversarial prompts) been tested?	Red team or independent assurance		
	Are lifecycle & readiness issues considered before deployment?	Lifecycle management plan, vendor maturity review		

Glossary

AI Agent An autonomous AI system capable of planning, sequencing, and executing tasks with minimal human intervention. Agents can access tools, browse the web, run code, and make decisions.

AI Car A metaphor used in this book to describe the full AI product / system - the visible experience powered by the underlying model

AI Engine A metaphor for the model itself. The underlying component that generates intelligence and output within an AI product.

AI Governance Framework The combination of policies, processes, and oversight structures that define how AI is adopted, monitored, and aligned with organizational strategy and ethics.

AI Journey The organizational process of defining purpose, developing strategy, setting governance, and managing risks in AI adoption.

AI Practices. These are the mechanics/maintenance of an AI Model (tuning, deployment, guardrails, oversight).

AI Safety Standard – A benchmark document (such as Australia's Voluntary AI Safety Standard 2024) outlining baseline for transparency, accountability, and oversight.

AI Systems the car that wraps the engine, adds controls, dashboard, and a driver experience (e.g., ChatGPT, Copilot, Perplexity).

Algorithm A set of rules or instructions that a computer follows to solve a problem or complete a task. In AI, algorithms process data to detect patterns, make predictions, or generate outputs.

Artificial Intelligence (AI) A broad field of computer science focused on building systems that can perform tasks that usually require human intelligence, such as understanding language, recognizing images, or making decisions.

Automation Repeatable set of instructions.

Bias (Algorithmic Bias) When an AI system produces unfair or skewed results because the data it was trained on reflects human or systemic prejudices. Example: facial recognition that performs worse on women or people of color.

Black Box A model whose internal decision-making process is opaque or not easily explained. Directors may see results without understanding how they were produced.

ChatGPT A consumer-facing application created by OpenAI in 2022. "Chat" refers to the interface (the text box users type into), while "GPT" refers to the underlying model, the Generative Pre-trained Transformer.

Cloud Computing The delivery of computing resources (servers, storage, networking, AI models) over the internet by providers like AWS, Microsoft Azure, or Google Cloud.

Control Effectiveness A measure of how well a policy, process, or safeguard mitigates risk. Often tracked through testing, scorecards, or audits.

Dark Data Information that an organization collects but never uses. It consumes storage, increases cost and carbon impact, and can pose privacy or compliance risks if mishandled.

Data Drift / Model Drift When an AI model's performance degrades over time because the data it sees in practice differs from the data it was trained on.

Data Governance The framework of policies and controls that ensures data is managed responsibly, securely, and in compliance with regulations.

Data Minimization A privacy and efficiency principle that limits data collection to only what is necessary for a defined purpose, reducing both exposure and storage waste.

Data Poisoning The intentional or accidental manipulation of training data to corrupt or bias an AI model's outputs.

Data Sovereignty The concept that data is subject to the laws and governance structures of the country in which it is collected or stored.

Deep Learning A subset of machine learning that uses multi-layered neural networks to process very large and complex datasets, powering image recognition, speech processing, and generative AI.

Edge Computing Processing data closer to where it is generated (e.g., on a local device or micro data center) rather than sending everything to a central cloud server.

Ethics (in AI) Principles guiding what is "right" or "just" in AI use, beyond what is merely legal. Examples include fairness, transparency, accountability, and respect for privacy.

Explainability The ability to describe in understandable terms how and why an AI system made a particular decision.

Explainable AI (XAI) AI systems designed to make their reasoning transparent and interpretable to humans, helping meet legal and ethical accountability standards.

Fiduciary Duty The legal responsibility of directors to act in good faith, with care, diligence, and in the best interests of the company and its shareholders.

Foundation Model A large-scale AI model (such as GPT-4, Claude, or Gemini) trained on massive datasets and adaptable to many downstream tasks.

Generative AI AI that does not just analyze existing data but creates new content — such as text, images, audio, or video — that mimics human work.

Governance (AI Governance) The oversight, policies, and accountability structures that ensure AI is deployed responsibly and aligned with organizational values and legal obligations.

Hallucination (AI Hallucination) When an AI confidently generates an answer that is false, misleading, or entirely fabricated.

Human-in-the-Loop (HITL) A safeguard that ensures human review and intervention in AI processes that affect customers, safety, or reputation.

Hybrid Cloud A computing environment that combines on-premises systems with private and public cloud services, often used for flexibility and redundancy.

Interpretability The ability to understand why and how a model makes a specific decision or how it produces an output.

Identity and Access Management (IAM) Systems and processes that control who can access which digital resources, and under what conditions.

Know Your Customer (KYC) is a regulatory process used by organizations to verify the identity of clients and assess potential risks of illegal activity such as money laundering or fraud. It involves collecting and validating identification documents, financial information, and customer intent before providing services.

Machine Learning (ML) A subset of AI where systems learn from data rather than being explicitly programmed. ML models improve their performance as they are exposed to more data.

Micro Data Centre A small, locally powered data center that supports specific regions or applications. Can improve resilience and reduce environmental impact compared to mega data centers.

Model (AI Model) The engine (e.g., GPT-4, Claude 3, Gemini Ultra) that takes data as input and produces an output (prediction, classification, or generation) based on patterns it has learned.

Model Governance A structured process that ensures models are designed, tested, approved, and monitored according to organizational and regulatory standards. It includes ownership, version control, explainability, and risk assessment.

Natural Language Processing (NLP) The field of AI focused on enabling machines to understand, interpret, and generate human language.

Neural Network A computing system inspired by the human brain, built from layers of "neurons" that process data in interconnected ways. Used heavily in deep learning.

Pilot (AI Pilot)
A limited trial of an AI system designed to test feasibility and risks before broad deployment.

Prompt The instruction or question given to a generative AI system that guides its output.

Prompt Engineering The practice of designing and refining inputs (prompts) to guide how generative AI systems respond. Effective prompt engineering improves accuracy and consistency, while poor or malicious prompts can cause bias, hallucinations, or data leakage.

Prompt Injection A form of attack where hidden or malicious instructions are inserted into prompts or data inputs, tricking an AI model into ignoring safety rules, revealing confidential information, or performing unintended actions.
Shared Responsibility Model In cloud computing, the principle that the provider secures the infrastructure, but the customer remains responsible for how data and applications are configured, accessed, and used.

Reinforcement Learning A learning method where an AI agent improves its decisions through trial and error, guided by rewards or penalties. Example: training a system to play chess or manage resource allocation by rewarding successful strategies and discouraging poor ones.

Responsible AI A collective term encompassing fairness, safety, inclusivity, and accountability in AI design and use.

Retrieval-Augmented Generation (RAG) A technique where generative AI systems retrieve relevant information from external sources before generating a response, improving accuracy but increasing exposure to prompt injection risks.

Residual Risk The level of risk that remains after all controls and mitigations have been applied. Boards should review whether residual risk aligns with the organization's risk appetite.

Roadmap A sequenced plan that outlines the steps, investments, and milestones for adopting AI. It turns strategy into action by showing when and how initiatives will be tested, scaled, and reviewed.

Shadow IT The use of unauthorized apps or tools by employees without approval from IT or security teams.

Software Supply Chain All the third-party code, components, and services that make up modern applications. A vulnerability in any part of the chain can affect the entire system.

Strategy A structured plan that defines why and how an organization adopts AI. It sets direction, measurable goals, and guardrails so adoption aligns with purpose, governance, and values, rather than drifting into risk or misuse.

Strategic Drift The gradual loss of alignment between an organization's AI strategy and its original purpose or values, often caused by unchecked growth or external pressure.

Supervised Learning A machine learning approach where models are trained on labeled data – meaning the input and the correct output are both provided. The model learns to map one to the other. Example: predicting whether a transaction is "fraud" or "legitimate" based on past labeled cases.

Tokenization The process of converting text into smaller units (tokens) such as words or characters, which the model can understand and process

Trust In governance terms, the social license to operate. Trust is earned when organizations act transparently, ethically, and responsibly, and lost when they act only to the letter of the law.

Unsupervised Learning A machine learning approach where models are given unlabeled data and must find hidden patterns or groupings on their own. Example: clustering customers with similar behaviors without being told the categories in advance.

Zero Trust A modern security principle that assumes no user or system is trusted by default. Every request for access must be verified, regardless of location or prior status.

Zero Trust Architecture (ZTA) A cybersecurity model that continuously verifies every user and device. When applied to AI, it limits model access, enforces authentication, and isolates systems to prevent propagation of compromise.

Index

A

B

O

P

Q

R

S

T

U

V

W

Z

Acknowledgement

Special Thanks to these INCREDIBLE people who have joined me on this journey to publishing this book.

Josh Taylor, Alex Donovan, James Harris, David Buitendag, Ben Edwards, Calvin Wang, Tony Johnston, Brendon Cappelletti, Aaron Teo, Louis Feyt, James Papadopoff, Dwight Jolly, Jake Spinner, Shauna Ganly, Alec Shaw, Warren Willis, Daisy St Claire, Saiyra Beri, Peter Markwell, Craign Geappen, Mitchell Young, Janice Vermeulen, Biljana Roksandić, Kelly Howell, Jishu Sen, Heide Fisher, Jeanette Hirniak-Morenski, Cara Mansfield, Karen Geappen, Mark Jackson, Eric Wong, Dinos Kerigan-Kyrou, Oliver Chikumbo, Dan Gregory, Vicki Gatewood, Mark Jackson, Micheal Jarvis, Aleks Krstic, Sheryl Gourdis, James Paterson, Peter Gigengack, Petar Lackovic, Kathleen Duncan, Barry Duncan, Tania McDonald, Suma Kamath, Morten Rosenkilda, Nathan Offringa, Peter Weetman, Karen Stewart, Jess Kaun, Julie Read, Kate Webb, Matthew Offringa, Richard Bremmer, Vicki Coles, Nicola Green, Tricia Kingston, Luisa Amanda, Drew Goodchild, EJ Wise, R. van Erven, Leone Kloppers, Charlotte Hanson, Robert Cheung, David Tabb, Tanya Pereira, Angela Newland, Makaita Chikoka, Varadhan Parthasarathy,Kathryn Jamieson, Dr Velu Ramasamy, Peta Demidenko, Tim Downing, Ruman Sarawer, Mike Schuman, Maria Duffy, Rebecca Chapman, Ryan Mills, Alexandre BLANC, Philip Savaete, Lorinda Gregoy, Caitriona Forde, Clarke Osborne, Adelia Neethling, Syed Ahmed, Damien Drake-Brockman, Craig Macaulay, Kaela Bonomi, Anna Holm, Max Vogt, Farrell Tirtadinata, Prof. Dan Haagman, Idris Kaka, Silvana M., Simon Carabetta, Gavin Rainbow, Karen Thomas, Jackson Chen, Barbara Prashnig, Azada Rudnicki, Daniel Piczak, Jen Li Wong, Andra Cimpean, Lovlin Rashmi, Niamh Toohey, Jeff Grbelja, Daniel Strydom, Evan Champion, Lisa Chan, Robert Jordan, Daniel Henderson, Lynda M., Ben Aylett, Nabila Farah, Yulia Merrill, Naeem Kyme, Carmen Maughen, Ryan Williams, Hazel Khoo, Aleisha Waters, Kathleen McGinley , Maddie Abe, Simon Thuijs, Tom Crijns, Andy Battle, Vanessa van Beek, Kanika Joshi, Adrian Bennet, Cecily Rawlinson, Eloise Pitts, Andreas Muller, Amanda Davidson, Odaya Amsellem, Joel Tan, Vishal Koul, Meidi van der Lee, Sameera Bandara, Janette Opperman, Maree Morrell, Joel O'Brien, Bill Perger, Andrew Dickson, Jeffrey Jay Blatt, Berkeley Baker, Preethi Koppisetti, Edwyn Tiew, Isaiah Imaseun, Wayne Ho, Vidhi Malkan, Paul Karan, Yusuf Denath, Dr Alan Briggs, Mohammed Huque, James Brine, Amanda Haley, Alex Fenna, Jeff Yeoman, Pene Quin, Esther & Ian Cowper, Lynda Carsons, Dr Stephen Shaw, Steve Yoo, Tammie Crawley, Ed Skoudis (and team) and ...[I am sure I left someone off this list].

From me to you

Thank you for sharing this adventure, for motivating me, and for encouraging me to finish this goal #grateful

Acknowledgement

[illegible]

About the Author

Brenda van Rensburg is a technology strategist, author, and international speaker who has spent more than two decades at the intersection of cybersecurity, governance, and artificial intelligence. She holds a Bachelor of Science, a Bachelor of Laws, and a Postgraduate in Artificial Intelligence, together with several preofessional industry qualifications.

Brenda has advised organizations across multiple sectors on strengthening digital resilience and aligning innovation with ethical and legal responsibility. Her insights have been featured in cybersecurity magazines and legal journals, where she continues to advocate for transparency, accountability, and trust in technology.

Driven by a lifelong curiosity about how machines think, and how humans can best use them, Brenda wrote On Board AI to help boards and leaders bridge the gap between technical complexity and responsible decision-making. She believes that the future of AI will be defined not by speed or scale, but by the integrity of those who lead it.

Made in United States
Orlando, FL
04 November 2026

90200391R00203